D0024199

INFINITE SERIES

INFINITE SERIES

JAMES M. HYSLOP

DOVER PUBLICATIONS, INC.
Mineola, New York

Bibliographical Note

This Dover edition, first published in 2006, is an unabridged republication of the fifth (1959) edition of the work, originally published in the "University Mathematical Texts" series by Oliver and Boyd, Edinburgh, and Interscience Publishers, Inc., New York, in 1942.

Library of Congress Cataloging-in-Publication Data

Hyslop, James M.
Infinite series / James M. Hyslop.
p. cm.
"This Dover edition, first published in 2006, is an unabridged republication of the fifth (1959) edition of the work, originally published in the "University Mathematical Texts" series by Oliver and Boyd, Edinburgh, and Interscience Publishers, Inc., New York, in 1942."—copyright p.
Includes index.
ISBN 0-486-45033-3 (pbk.)
1. Series, Infinite. I. Title.

QA295.H9 2006
515'.243—dc22

2005054783

Manufactured in the United States of America
Dover Publications, Inc., 31 East 2nd Street, Mineola, N.Y. 11501

PREFACE

THIS book is intended for second- or third-year students who have some knowledge of the principles of elementary analysis. Definitions of the terms and summaries of those results in analysis which are of special importance in the theory of series are given in Chapter I. Where it has proved convenient the o, O notation has been used, even although this is sometimes considered too difficult for the average student. In the interests of rigidity it has been necessary to discuss the question of the upper and lower limits of a function, but I have confined myself to an outline of those properties which have direct bearing on the convergence of series.

The central theme of the book is the convergence of *real* series, but series whose terms are complex and real infinite products are also discussed as illustrations of the main theme. Infinite integrals have been omitted, except in connection with the integral test for convergence.

In an elementary book of this kind it is difficult to state, with any accuracy, to whom I am indebted for the particular presentation of the subject, but the lecturers of my student days, Professor T. M. MacRobert, Dr James Hyslop and Mr A. S. Besicovitch, must have influenced me considerably. I am especially indebted to Professor MacRobert, who, in view of my absence from home, has very kindly corrected all the proofs for me. My thanks are also due to Dr Graham, who has verified the examples, and to Dr Rutherford, who, along with Professor MacRobert, has seen the book through the press.

J. M. HYSLOP

R.A.F.
Middle East
August 1942

PREFACE TO FIFTH EDITION

In this edition a few errors have been corrected and a few minor alterations have been made in the text, in the interests of clarity.

J. M. HYSLOP

April 1954

CONTENTS

CHAPTER I

FUNCTIONS AND LIMITS

		PAGE
1.	Introduction	1
2.	Functions	1
3.	Bounds of a Function	2
4.	Limits of Functions	3
5.	Two Important Limits	6
6.	Monotonic Functions	8
7.	Upper and Lower Limits	8
8.	Continuity	10
9.	Differentiation	11
10.	Integration	12
11.	The o, O notation	14
	Examples	15

CHAPTER II

SOME PROPERTIES OF PARTICULAR FUNCTIONS

12.	The Logarithmic and Exponential Functions	18
13.	The Hyperbolic Functions	22
14.	The Circular Functions	23
	Examples	23

CHAPTER III

REAL SEQUENCES AND SERIES

15.	Definition of a Sequence	25
16.	Convergent, Divergent and Oscillating Sequences	25
17.	Infinite Series	26
18.	Important Particular Series	26
19.	The General Principle of Convergence	29
20.	Some Preliminary Theorems on Series	31
	Examples	33

CHAPTER IV

SERIES OF NON-NEGATIVE TERMS

PAGE
21. A Fundamental Theorem 36
22. Rearrangement of Terms 36
23. Tests for Convergence 38
24. The Integral Test 38
25. The Comparison Tests 40
26. The Ratio or d'Alembert's Test 43
27. Cauchy's Test 44
28. Connection between the Ratio Test and Cauchy's Test . 45
29. A General Test for Convergence 46
30. Raabe's Test 48
31. Gauss's Test 48
32. Euler's Constant 51
33. Stirling's Approximation for $n!$ 51
Examples 54

CHAPTER V

GENERAL SERIES

34. Real Series 58
35. Absolute Convergence 58
36. Tests for Absolute Convergence 59
37. Conditional Convergence 60
38. Riemann's Theorem 62
39. Complex Limits 63
40. Series whose Terms may be Complex 64
41. Abel's Lemma 65
Examples 66

CHAPTER VI

SERIES OF FUNCTIONS

42. Uniform Convergence 68
43. Series of Functions 69
44. Tests for Uniform Convergence 70
45. Some Properties of Uniformly Convergent Series . . 73
46. Power Series 79
Examples. 83

CHAPTER VII

THE MULTIPLICATION OF SERIES

PAGE

47. Multiplication of Series of Non-Negative Terms . . 86
48. Multiplication of General Series 87
    Examples 91

CHAPTER VIII

INFINITE PRODUCTS

49. Convergence and Divergence of Infinite Products . . 93
50. Some Theorems on Special Types of Products . . 94
51. The Absolute Convergence of Infinite Products . . 96
52. The Uniform Convergence of an Infinite Product . . 97
53. The Infinite Products for sin x and cos x 98
54. The Gamma Function 102
    Examples 105

CHAPTER IX

DOUBLE SERIES

55. Introduction 108
56. Double Series whose Terms are Non-negative . . 110
57. The Absolute Convergence of a Double Series . . 113
58. The Interchange of the Order of Summation for Repeated Series 115
    Examples 117

Index 119

CHAPTER I

FUNCTIONS AND LIMITS

1. Introduction. The theory of infinite series is an important branch of elementary mathematical analysis. For its proper understanding it is essential for the reader to have some knowledge of such fundamental ideas as bounds, limits, continuity, derivatives and integrals of functions. In this chapter a brief sketch will be given of those results in analysis which will be used in the book, and also a more detailed discussion of the question of limits. It will be assumed that the reader is familiar with the simple properties of the logarithmic, exponential, hyperbolic and circular functions. Certain properties of these functions, however, which are of special importance in the theory of series will be derived in Art. **18**.

2. Functions. It is sufficient for our purpose to regard a function of a variable as a mathematical expression which possesses one calculable value corresponding to each of a set of values of the variable. Each calculated value of the expression is called the value of the function corresponding to the appropriate value of the variable. Throughout, the letter x or y will denote a **real** variable, that is, a variable which takes only real values and, unless otherwise stated, the functions with which we deal will also be assumed to be real, that is, to possess only real values. Functions of x are usually denoted by symbols such as $F(x)$, $f(x)$, $\phi(x)$, etc., and their values when $x = a$ by $F(a)$, $f(a)$, $\phi(a)$, etc. If values of the function $f(x)$ can be determined for certain values of the variable x we say that $f(x)$ is **defined** for these values of x. If the function $f(x)$ is defined for all values of x satisfying the inequality $a<x<b$, we say that $f(x)$ is defined in the **open interval** (a, b). If, in addition, $f(x)$ is defined for $x = a$ and for $x = b$, then $f(x)$ is defined for $a \leqslant x \leqslant b$ and we say that $f(x)$ is defined in the **closed interval** (a, b).

3. Bounds of a Function. Suppose that the function $f(x)$ is defined for a certain set of values of x. If there is a number which is greater than all the values of $f(x)$ then $f(x)$ is said to be **bounded above** for these values of x. If there is a number which is smaller than all the values of $f(x)$ then $f(x)$ is said to be **bounded below** for these values of x. If both conditions are satisfied $f(x)$ is said to be **bounded** for these values of x.

If, for a certain set of values of x, there is a number K, independent of x, such that (i) $f(x) \leqslant K$, (ii) there is at least one value of x for which $f(x)>K-\epsilon$, where ϵ is any positive number,* then K is called the **upper bound** of $f(x)$ for this set of values of x. If, for a certain set of values of x, there is a number k, independent of x, such that (i) $f(x) \geqslant k$, (ii) there is at least one value of x for which $f(x)<k+\epsilon$, then k is called the **lower bound** of $f(x)$ for this set of values of x.

It is clear that the lower bound of $f(x)$ is not greater than its upper bound.

The functions

$$\tan x, \ (0 \leqslant x < \tfrac{1}{2}\pi),$$
$$(-1)^n n, \ (n \text{ a positive integer}),$$
$$\sin (1/x), \ (x>0),$$

are, respectively, unbounded above and bounded below with lower bound zero, unbounded above and below, bounded above and below with upper and lower bounds $+1$ and -1.

* Throughout ϵ and η will always denote *positive* numbers and it is convenient to think of them as small.

The following theorem is fundamental.*

THEOREM A. *If $f(x)$ is bounded above for a certain set of values of x, it possesses an upper bound for these values of x. If $f(x)$ is bounded below it possesses a lower bound.*

4. Limits of Functions. The function $f(x)$ is said to tend to the limit l as x tends to a if, given ϵ, we can † find $\eta = \eta(\epsilon)$ such ‡ that $|f(x)-l|<\epsilon$ for all values of x except a for which the function is defined and which also satisfy the inequality $|x-a|<\eta$. In these circumstances we write $f(x) \to l$ as $x \to a$ or $\lim_{x \to a} f(x) = l$.

The function $f(x)$ is said to tend to the limit l as x tends to infinity if, given ϵ, we can find $X = X(\epsilon)$ such that $|f(x)-l|<\epsilon$ for all values of $x>X$ for which the function is defined. In these circumstances we write $f(x) \to l$ as $x \to \infty$ or $\lim_{x \to \infty} f(x) = l$.

The function $f(x)$ is said to tend to infinity as x tends to infinity if, given any positive number K, we can find $X = X(K)$ such that $f(x)>K$ for all values of $x>X$ for which the function is defined. In these circumstances we write $f(x) \to \infty$ as $x \to \infty$ or $\lim_{x \to \infty} f(x) = \infty$.

The reader should also construct definitions corresponding to the expressions

$$\lim_{x \to a} f(x) = -\infty, \quad \lim_{x \to -\infty} f(x) = l, \quad \lim_{x \to -\infty} f(x) = \infty.$$

Throughout the remainder of this article we shall consider only limits as $x \to \infty$ and we shall assume that

* Theorems A, B, C are stated without proof. Their proofs will be found in most text-books on mathematical analysis.

† The symbol $|x|$ means the numerical value of x. For example, $|3| = 3$, $|-2| = 2$.

The inequalities $|x_1+x_2| \leqslant |x_1|+|x_2|$, $|x_1 \pm x_2| \geqslant |x_1|-|x_2|$ are easy to verify.

‡ The statement $\eta = \eta(\epsilon)$ means η depending only on ϵ.

the functions under discussion are defined for all sufficiently large values of x. There being no possibility of ambiguity we shall use the contracted notation $\lim f(x)$ for a limit of this kind. The subsequent theorems hold with trivial modifications for other types of limits and, in particular, for limits as $x \to \infty$ through a certain set of values. It will be observed that from the definition of a limit it follows that, if $f(x)$ is defined for all values of x and if $\lim f(x) = l$, then *a fortiori* $f(x) \to l$ as x tends to infinity through any set of values and, in particular, through all positive integral values. We now prove some fundamental theorems on limits.

THEOREM 1. *If* $\lim f_1(x) = l_1$, $\lim f_2(x) = l_2$, *then*

(i) $\lim \{f_1(x)+f_2(x)\} = l_1+l_2$,

(ii) $\lim f_1(x)f_2(x) = l_1 l_2$,

(iii) $\lim f_1(x)/f_2(x) = l_1/l_2$,

where, in (iii), $l_2 \neq 0$.

Corresponding to any positive number θ we can find $X_1 = X_1(\theta)$, $X_2 = X_2(\theta)$ such that

$$|f_1(x)-l_1|<\theta\,,\ |f_2(x)-l_2|<\theta,$$

whenever $x>X_1$, $x>X_2$ respectively. If $X = \text{Max}\,(X_1, X_2)$, that is, if X is the larger of X_1 and X_2, then these two inequalities hold *a fortiori* whenever $x>X$.

(i) Given ϵ, let $\theta = \frac{1}{2}\epsilon$ and determine X as above. Then whenever $x>X$, which depends only on ϵ,

$$\begin{aligned} |f_1(x)+f_2(x)-l_1-l_2| &\leqslant |f_1(x)-l_1|+|f_2(x)-l_2| \\ &< 2\theta \\ &= \epsilon, \end{aligned}$$

and this proves (i).

(ii) Given ϵ, let θ be the positive root of the equation

$$x^2+(|l_1|+|l_2|)x-\epsilon = 0,$$

and determine X as a function of θ, and therefore of ϵ, as before. Then, whenever $x > X$,

$$\begin{aligned}|f_1(x)f_2(x)-l_1l_2| &= |f_1(x)\{f_2(x)-l_2\}+l_2\{f_1(x)-l_1\}| \\ &\leqslant |f_1(x)|\ |f_2(x)-l_2|+|l_2|\ |f_1(x)-l_1| \\ &< (|l_1|+\theta)\theta+|l_2|\theta \\ &= \theta^2+(|l_1|+|l_2|)\theta \\ &= \epsilon,\end{aligned}$$

which proves (ii).

(iii) Given ϵ, let θ be any positive number satisfying both the inequalities

$$\theta \leqslant \tfrac{1}{2}|l_2|,\ \theta \leqslant \frac{\epsilon l_2{}^2}{2|l_1|+4|l_2|},$$

and determine X as before. Then,* whenever $x > X$,

$$\begin{aligned}\left|\frac{f_1(x)}{f_2(x)}-\frac{l_1}{l_2}\right| &= \left|\frac{l_2f_1(x)-l_1f_2(x)}{l_2f_2(x)}\right| \\ &= \left|\frac{f_1(x)\{l_2-f_2(x)\}+f_2(x)\{f_1(x)-l_1\}}{l_2f_2(x)}\right| \\ &\leqslant \frac{|f_1(x)|\,|l_2-f_2(x)|+|f_2(x)|\,|f_1(x)-l_1|}{|l_2|\,|f_2(x)|} \\ &< \frac{(|l_1|+\theta)\theta+(|l_2|+\theta)\theta}{|l_2|(|l_2|-\theta)} \\ &\leqslant \theta\left\{\frac{|l_1|+\frac{1}{2}|l_2|+|l_2|+\frac{1}{2}|l_2|}{\frac{1}{2}l_2{}^2}\right\} \\ &\leqslant \epsilon,\end{aligned}$$

which proves (iii).

It should be noted that (i) and (ii) hold not merely for two but for any finite number of functions $f_1(x)$, $f_2(x)$, $f_3(x)$, . . . The reader should examine how far the theorem remains true in the case when either l_1 or l_2 or both are infinite.

* It is assumed here that $f_2(x) \neq 0$ for any particular value of x. If it is zero we merely omit the corresponding value of x from consideration.

THEOREM 2. *If, for all sufficiently large values of x, we have $f_1(x) \leqslant f_2(x)$ then $l_1 \leqslant l_2$, where $l_1 = \lim f_1(x)$ and $l_2 = \lim f_2(x)$ and it is assumed that these limits exist.*

Suppose if possible that $l_1 > l_2$. Let ϵ be $\frac{1}{2}(l_1 - l_2)$. Then, as in the proof of Theorem 1, we can determine $X = X(\epsilon)$ such that, whenever $x > X$,

$$l_1 - \epsilon < f_1(x) < l_1 + \epsilon, \quad l_2 - \epsilon < f_2(x) < l_2 + \epsilon.$$

It follows that, for such values of x,

$$f_1(x) - f_2(x) > l_1 - \epsilon - l_2 - \epsilon = 0,$$

which contradicts the hypothesis.

The theorem is therefore proved.

COROLLARY. *Under the conditions of Theorem 2 if* $\lim f_1(x) = \infty$, *then* $\lim f_2(x) = \infty$.

It should be noted in passing that the hypothesis $f_1(x) < f_2(x)$ does not imply the conclusion $l_1 < l_2$. For example, if $f_1(x) = x^{-2}$, $f_2(x) = x^{-1}$, then $f_1(x) < f_2(x)$ when $x > 1$ but $\lim f_1(x) = \lim f_2(x) = 0$.

5. Two Important Limits. Besides being of great importance in themselves the limits which we shall now discuss serve to illustrate the preceding definitions and theorems.

(i) Let $f(n) = n^a x^n$, where n is a positive integer * and a and x are any real numbers.† We shall determine $\lim_{n \to \infty} f(n)$ for all values of a and x.

Suppose first that $|x| < 1$, $a \geqslant 0$. Let ‡ $p = [a] + 1$. If $x = 0$, $f(n) = 0$ for all values of n so that $\lim f(n) = 0$.

* Throughout the book the letters n, m, p will always represent a positive integer (or zero).

† The definition of the symbol n^a for all real values of a will be given in Art. 12.

‡ The symbol $[x]$ means the integral part of x. For example $[\frac{5}{2}] = 2$, $[3] = 3$.

If $x \neq 0$ we may write $|x| = 1/(1+a)$, where $a > 0$. Then, for $n > p+1$,

$$|f(n)| \leqslant \frac{n^p}{(1+a)^n} = \frac{n^p}{1+na+\binom{n}{2}a^2+\ldots+a^n}$$

$$< \frac{n^p}{\dfrac{n(n-1)\ldots(n-p)}{1.2\ldots(p+1)}a^{p+1}}$$

$$= \frac{(p+1)!\, a^{-p-1}n^{-1}}{(1-\frac{1}{n})(1-\frac{2}{n})\ldots(1-\frac{p}{n})}$$

$$\to 0,$$

as $n \to \infty$, since the numerator tends to zero and the denominator is a product of p factors each of which tends to 1. Thus, by Theorem 2, $\lim f(n) = 0$. If $|x| < 1$, $\alpha < 0$ the same result is true, since

$$n^\alpha |x|^n < |x|^n \to 0.$$

Suppose now that $|x| > 1$ and $\alpha \leqslant 0$. Write $\beta = -\alpha$ and $y = 1/x$. Then

$$|f(n)| = \frac{1}{n^\beta |y|^n} \to \infty,$$

as $n \to \infty$, since $\beta \geqslant 0$ and $|y| < 1$. If $|x| > 1$, $\alpha > 0$ the same result is true, since

$$n^\alpha |x|^n > |x|^n \to \infty.$$

Thus, when $x > 1$, $f(n) \to \infty$. When $x < -1$, $f(n) \to \infty$ as n tends to infinity through even values and $f(n) \to -\infty$ as n tends to infinity through odd values. The function $f(n)$ is therefore unbounded when $|x| > 1$.

When $x = 1$ we have $f(n) = n^\alpha$ and $\lim f(n) = 1, \infty, 0$ according as $\alpha = 0, > 0, < 0$.

When $x = -1$, $\alpha \geqslant 0$, we have $f(n) = (-1)^n n^\alpha$, which does not tend to a limit as n tends to infinity. In the case $x=-1$, $a=0$, however, $f(n)$ is bounded and has upper and lower bounds equal to $+1$ and -1. When $x=-1$, $a<0$ the function clearly tends to zero.

(ii) Let $\phi(n)=x^n/n!$ We shall show that, for all values of x, $\phi(n)$ tends to zero as n tends to infinity.

Let $N=[|x|]$. Then, if $n>N$,

$$\left|\frac{x^n}{n!}\right|=\frac{|x|^N|x|^{n-N}}{N!(N+1)(N+2)\ldots n}<\frac{|x|^N}{N!}\left(\frac{|x|}{N+1}\right)^{n-N}\to 0,$$

as $n\to\infty$, by (i), since $0\leqslant|x|<N+1$. The result follows from Theorem 2.

6. Monotonic Functions. If, as x increases in a certain interval (a, b), the function $f(x)$ does not decrease, then $f(x)$ is called a **monotonic increasing** function of x in (a, b); if $f(x)$ does not increase, then it is called a **monotonic decreasing** function of x in (a, b).

THEOREM 3. *If $f(x)$ is a monotonic increasing (decreasing) function of x for $x>a$ then, as $x\to\infty$, $f(x)$ tends to a finite limit or to $+\infty(-\infty)$ according as $f(x)$ is bounded above (below) or not.*

It will be sufficient to prove the theorem for a monotonic increasing function only.

Suppose that $f(x)$ is bounded. By Theorem A it has an upper bound K with the properties, (i) $f(x)\leqslant K$ if $x>a$; (ii) given ϵ, there is a value X of x greater than a such that $f(X)>K-\epsilon$. Since $f(x)$ is monotonic increasing it follows that, whenever $x\geqslant X$,

$$K-\epsilon<f(x)\leqslant K<K+\epsilon.$$

Thus $f(x)\to K$ as $x\to\infty$.

Suppose that $f(x)$ is not bounded above. Given any positive number L we can find a value X' of x such that $f(X')>L$. It follows that $f(x)>L$ for all values of $x\geqslant X'$. Hence $f(x)\to\infty$ as $x\to\infty$.

7. Upper and Lower Limits. Suppose that the function $f(x)$ is bounded for all values of $x\geqslant x_0$. Let

$M(X)$, $m(X)$ denote respectively the upper and lower bounds of $f(x)$ for $x \geqslant X \geqslant x_0$. Then $M(X)$, $m(X)$ are respectively monotonic decreasing and monotonic increasing bounded functions of X. By Theorem 3 they therefore tend to finite limits as $X \to \infty$. These limits are called respectively the upper and lower limits of $f(x)$ as $x \to \infty$, and we write

$$\overline{\lim} f(x) = \overline{\lim_{x \to \infty}} f(x) = \lim_{x \to \infty} M(X),$$

$$\underline{\lim} f(x) = \underline{\lim_{x \to \infty}} f(x) = \lim_{x \to \infty} m(X).$$

It should be observed that every bounded function possesses finite upper and lower limits and that the lower limit is not greater than the upper limit.

As an illustrative example consider the following function of the positive integral variable n,

$$f(n) = 1 + (-1)^n + \frac{1}{n}.$$

If N is even, $M(N) = 2 + \frac{1}{N}$, $m(N) = 0$, while, if N is odd,

$$M(N) = 2 + \frac{1}{N+1}, \quad m(N) = 0.$$

Thus $\qquad \overline{\lim} f(n) = 2, \quad \underline{\lim} f(n) = 0.$

It is clear that

$$\overline{\lim} \sin x = 1, \underline{\lim} \sin x = -1.$$

The following theorem gives us an alternative definition for upper and lower limits.

THEOREM 4. *If there is a number l such that,* (i) *given ϵ there exists $X_1 = X_1(\epsilon)$ such that $f(x) < l + \epsilon$ whenever $x \geqslant X_1$,* (ii) *no matter how large X_2 may be there is a value of $x \geqslant X_2$ for which $f(x) > l - \epsilon$, then $l = \overline{\lim} f(x)$. Conversely, if $\overline{\lim} f(x) = l$, then l has the properties* (i) *and* (ii).

Similar properties hold in the case of the lower limit.

Since $M(X_1)$ is the upper bound of $f(x)$ for $x \geqslant X_1$, there is a number $X'_1 \geqslant X_1$, such that $f(X'_1) > M(X_1) - \epsilon$. Thus, whenever $X \geqslant X_1$,

$$M(X) \leqslant M(X_1) < f(X'_1) + \epsilon < l + 2\epsilon.$$

Also, from (ii), we have $M(X) > l - \epsilon$. Hence $M(X) \to l$ as $X \to \infty$; that is, $l = \overline{\lim} f(x)$.

Conversely, if $\overline{\lim} f(x) = l$, given ϵ, there is a number $X_1 = X_1(\epsilon)$ such that $M(X) < l + \epsilon$ whenever $X \geqslant X_1$. In particular, $M(X_1) < l + \epsilon$; whence $f(x) < l + \epsilon$ whenever $x \geqslant X_1$. Also there is a number X_2 such that $M(X_2) > l - \frac{1}{2}\epsilon$. Since $M(X_2)$ is the upper bound of $f(x)$ for $x \geqslant X_2$, there is a value of $x \geqslant X_2$ such that $f(x) > M(X_2) - \frac{1}{2}\epsilon$. For this value of x we then have $f(x) > l - \epsilon$.

For the case of the lower limit a similar proof may be constructed.

THEOREM 5. *If* $\overline{\lim} f(x) = \underline{\lim} f(x) = l$, *then* $\lim f(x) = l$, *and conversely.*

By Theorem 4, given ϵ we can find $X_1 = X_1(\epsilon)$ such that $f(x) < l + \epsilon$ whenever $x \geqslant X_1$ and $X_2 = X_2(\epsilon)$ such that $f(x) > l - \epsilon$ whenever $x \geqslant X_2$. Let $X = \text{Max}\,(X_1, X_2)$. Then $|f(x) - l| < \epsilon$ whenever $x \geqslant X$; that is, $\lim f(x) = l$.

We leave the proof of the converse to the reader.

8. Continuity. Suppose that the function $f(x)$ is defined in the interval $a \leqslant x \leqslant b$ and that x_0 is some point * (other than a or b) in this interval. Then $f(x)$ is said to be **continuous** at the point x_0 if $\lim\limits_{x \to x_0} f(x) = f(x_0)$. It is said to be continuous at a if $f(x) \to f(a)$ as x tends to a **from the right** (that is, through values of x greater than a), and at b if $f(x) \to f(b)$ as x tends to b **from the left** (that is, through values of x less than b). The function is said to be continuous in the interval $a \leqslant x \leqslant b$ if it is continuous at every point of the interval.

* Here and occasionally elsewhere it is convenient to use the language of geometry.

For example, the function x^{-1} is continuous for $x>0$ or for $x<0$ but not for $x = 0$ since it is not defined at this point. Also the function $f(x) = x$; $(x \neq 0)$, $f(0) = 1$ is not continuous at $x = 0$ since $\lim_{x \to 0} f(x) = 0 \neq f(0)$.

The following theorem summarises those properties of continuous functions which we require.

THEOREM B. *The sum, difference and product of two functions $f(x)$ and $\phi(x)$ which are continuous at x_0 are also continuous at x_0. Also $f(x)/\phi(x)$ is continuous at x_0 provided that $\phi(x_0) \neq 0$.*

9. Differentiation. If the function $f(x)$ is defined in the interval (a, b), and if x is a point in this interval, then

$$\lim_{h \to 0} \frac{f(x+h)-f(x)}{h},$$

if it exists, is called the **derivative** or **differential coefficient** of $f(x)$ at the point x, and we denote it by $f'(x)$ or by $\frac{d}{dx} f(x)$. The function $f(x)$ is then said to be differentiable at the point x. The derivatives of $f(x)$ at a and b are defined similarly with the same conventions in regard to the limit operations as in the case of continuity. If the above limit exists for all points x in the interval (a, b) then $f(x)$ is said to be differentiable in the interval. Its derivative $f'(x)$ is then defined for all points in the interval (a, b). If $f'(x)$ is differentiable in (a, b) we denote its derivative by $f''(x)$. Similarly, we may obtain in succession further higher derivatives of $f(x)$.

We assume that the reader is familiar with the ordinary rules of differentiation and with the derivatives of functions which commonly occur in elementary analysis. We state the following theorem for the purpose of reference.

THEOREM C. (i) *If $f'(x_0)$ exists then $f(x)$ is continuous at the point x_0.* (ii) *If the n-th derivative $f^{(n)}(x)$ of the function*

$f(x)$ exists in an interval which includes the origin and if x is any point in this interval, then we have

$$f(x) = f(0) + \frac{x}{1!} f'(0) + \ldots + \frac{x^{n-1}}{(n-1)!} f^{(n-1)}(0) + \frac{x^n}{n!} f^{(n)}(\theta x),$$

or

$$f(x) = f(0) + \frac{x}{1!} f'(0) + \ldots + \frac{x^{n-1}}{(n-1)!} f^{(n-1)}(0) + \frac{x^n}{(n-1)!} (1-\theta)^{n-1} f^{(n)}(\theta x),$$

*where, in both cases,** $0<\theta<1$.

The expansions in (ii) are called the Taylor (or Maclaurin) expansions of $f(x)$.

10. Integration. For the purposes of this book it is not necessary for the reader to be acquainted with the strictly arithmetical definition of the definite integral. He should, however, know the " standard " integrals, the more important theoretical properties of integrals and the various methods of simplifying and evaluating them.

Such properties, together with a complete discussion of the arithmetical definition of the Riemann integral, will be found in R. P. Gillespie, *Integration*.†

Here we shall discuss briefly the question of infinite integrals, as these have important applications to the theory of series. In passing it is perhaps worth remarking that almost every theorem for infinite series has an exact analogue for infinite integrals.

If, for $t \geqslant a$, the function $f(t)$ is continuous it is known ‡ that $\int_a^x f(t)dt$ exists for $x \geqslant a$ and defines a function $F(x)$ which is continuous for $x \geqslant a$. If $\lim_{x\to\infty} F(x)$ is finite and equal to L the integral $\int_a^\infty f(t)dt$ is said to be **convergent** to

* It should be noted that θ depends both on n and on x.

† This text-book will be referred to as *G*. References are to the third edition. ‡ See *G*., pp. 74, 78.

the value L. Otherwise the integral is said to be **divergent.** Divergent integrals are further classified into **properly divergent, finitely oscillating** and **infinitely oscillating** integrals. In the first of these $F(x)$ tends to $+\infty$ or to $-\infty$, in the second $F(x)$ does not tend to a limit but remains bounded for all large values of x, while, in the third, $F(x)$ does not tend to a limit and is not bounded. For example, the integral $\int_a^\infty t^{-\lambda}dt$, $(a>0)$, is convergent for $\lambda>1$ and properly divergent for $\lambda\leqslant 1$, the integral $\int_0^\infty \sin t\, dt$ oscillates finitely and the integral $\int_0^\infty t \sin t\, dt$ oscillates infinitely.

Suppose that, for $t\geqslant a$, $f(t)$ and $g(t)$ are positive continuous functions and that $f(t)\leqslant g(t)$. Suppose further that $\int_a^\infty g(t)dt$ converges to the value M. Under these conditions it follows that $\int_a^\infty f(t)dt$ is convergent, for $F(x)$ is a monotonic increasing function of x and

$$F(x)\leqslant\int_a^x g(t)dt\leqslant M.$$

Hence, by Theorem 3, $F(x)$ tends to a finite limit.

Suppose now that $f(t)$ is not defined at the point a and that, elsewhere in the range $a\leqslant t\leqslant b$, $f(t)$ is continuous. In these circumstances $\int_a^b f(t)dt$ is *defined* to be $\lim_{\epsilon\to 0}\int_{a+\epsilon}^b f(t)dt$, and is said to be convergent if this limit is finite and divergent otherwise. Such an integral is essentially the same as that which we have already discussed, for it reduces to that type by employing a suitable change of variable. The comparison property which we have obtained above clearly holds also for this type of integral. We may similarly form definitions for $\int_a^b f(t)dt$ when $f(t)$ is not defined at b or when $f(t)$ is not defined at some point within the interval (a, b).

Example. Show that the integral $\int_0^{\frac{1}{2}\pi} \log (1/\sin \theta) d\theta$ is convergent.

We use the inequality $\sin\theta \geqslant \frac{2\theta}{\pi}$, $(0 \leqslant \theta \leqslant \frac{1}{2}\pi)$, and some elementary properties of the logarithmic function (see Art. **12**).

The integrand is positive and continuous for $0 < \theta \leqslant \frac{1}{2}\pi$ and is not defined at $\theta = 0$. We have

$$\int_0^{\frac{1}{2}\pi} \log (1/\sin\theta) d\theta \leqslant \int_0^{\frac{1}{2}\pi} \log (\pi/2\theta) d\theta$$
$$= \Big[\theta \log (\pi/2\theta)\Big]_{\theta\to 0}^{\frac{1}{2}\pi} + \int_0^{\frac{1}{2}\pi} d\theta,$$

which is finite. It therefore follows that the given integral is convergent.

The given integral is equal to $-\int_0^{\frac{1}{2}\pi} \log \sin \theta \, d\theta$ and its value will be found in Art. **33**.

11. The o, O notation. Let $\phi(x)$ be a positive function of x, that is, a function of x which takes only positive values, and let $f(x)$ be a second function which is defined for the same values of x as $\phi(x)$. If, for these values of x, there is a positive number K, independent of x, such that

$$|f(x)| < K\phi(x),$$

then we write $f(x) = O\{\phi(x)\}$. For example,

$$\sin x = O(|x|), \ (-\tfrac{1}{2}\pi \leqslant x \leqslant \tfrac{1}{2}\pi),$$
$$\cos x = O(1), \text{ for all values of } x.$$

The first relation is of course true for values of x outside the range $(-\frac{1}{2}\pi, \frac{1}{2}\pi)$, but, in such cases, the relation $\sin x = O(1)$ is more precise. By the statement $f(x) = O\{\phi(x)\}$ as $x\to\infty$ we mean that $f(x) = O\{\phi(x)\}$ for all values of x concerned which are greater than some fixed number. In the same way a meaning may also be attached to the statement $f(x) = O\{\phi(x)\}$ as $x \to a$.

If, as $x \to a$, $f(x)/\phi(x) \to 0$, then we write $f(x) = o\{\phi(x)\}$ as $x \to a$. For example,

$$\tan x^3 = o(x^2) \text{ as } x \to 0,$$
$$\sqrt{x} = o(x) \text{ as } x \to \infty, \quad x = o(\sqrt{x}) \text{ as } x \to 0.$$

It is often convenient to use symbols like $O(x)$, $o(1)$, $O(x^2)$, $o(x^{-1})$, etc., without reference to a specific function. For example, the symbol $O(x^2)$ stands for any function whose numerical or absolute value when divided by x^2 is bounded for the values of x under consideration. Again, the symbol $o(1)$ stands for any function which tends to zero as the variable under consideration tends to some number which is rendered unambiguous by the context. Meanings are thus attached to such statements as

$$O(1) = o(x) \text{ as } x \to \infty, \quad o(x) = o(\sqrt{x}) \text{ as } x \to 0.$$

If $f(x)$ and $\phi(x)$ are any two functions such that $f(x)/\phi(x) \to 1$ as $x \to a$ then we write $f(x) \sim \phi(x)$ as $x \to a$. For example, $\tan x \sim x$ as $x \to 0$.

Example. If, as $x \to \infty$, $f(x) = x^2 + O(x)$, $\phi(x) \sim x^{-1}$, show that $f(x)\phi(x) = x + o(x)$.

Since $\phi(x) \sim x^{-1}$ we may write $\phi(x) = x^{-1} + o(x^{-1})$. Then

$$\begin{aligned} f(x)\phi(x) &= \{x^2 + O(x)\}\{x^{-1} + o(x^{-1})\} \\ &= x + O(1) + o(x) + O(x) \,.\, o(x^{-1}). \end{aligned}$$

As $x \to \infty$, $O(1) = o(x)$ and $O(x) \,.\, o(x^{-1}) = o\,(1) = o(x)$. The result therefore follows.

Examples

1. Evaluate the following limits :—

(i) $\lim\limits_{n \to \infty} \dfrac{n^2 - n + 1}{2n^2 + 2n + 1}$,

(ii) $\lim\limits_{n \to \infty} \dfrac{a_0 n^p + a_1 n^{p-1} + \ . \ . \ . \ + a_p}{b_0 n^p + b_1 n^{p-1} + \ . \ . \ . \ + b_p}$, $(b_0 \neq 0)$,

(iii) $\lim\limits_{n \to \infty} \dfrac{x^n + n}{x^{n-1} + 2n}$,

(iv) $\lim_{n\to\infty} \dfrac{x-x^{2n+1}}{1+x^{2n+2}}$,

(v) $\lim_{x\to 0} \dfrac{\sqrt{(1+x)}-\sqrt{(1-x)}}{\sqrt{(2+x)}-\sqrt{(2-x)}}$.

2. If $\lim f(n) = l$ prove that

$$\lim \frac{f(1)+f(2)+\;.\;.\;.\;+f(n)}{n} = l.$$

[Write $f(n) = l+\phi(n)$. Then $\phi(n)\to 0$ as $n\to\infty$ and the result will be proved if we show that

$$\Phi(n) \equiv \frac{\phi(1)+\phi(2)+\;.\;.\;.\;+\phi(n)}{n} \to 0.$$

Given ϵ, we can find $N = N(\epsilon)$ such that $|\phi(n)|<\frac{1}{2}\epsilon$, whenever $n>N$. Writing

$$|\phi(1)+\phi(2)+\ldots+\phi(N)| = K,$$

we then have

$$\begin{aligned}|\Phi(n)| &\leqslant \frac{K}{n}+\frac{|\phi(N+1)|+|\phi(N+2)|+\ldots+|\phi(n)|}{n}\\ &< \frac{K}{n}+\frac{\epsilon(n-N)}{2n}\\ &< \frac{K}{n}+\tfrac{1}{2}\epsilon\\ &< \tfrac{1}{2}\epsilon+\tfrac{1}{2}\epsilon\\ &= \epsilon,\end{aligned}$$

whenever $n>$ Max. $(N, 2K/\epsilon)$. The result follows.]

3. If

$$f(n) = \frac{1}{\sqrt{\{n(n+1)\}}}+\frac{1}{\sqrt{\{(n+1)(n+2)\}}}+\ldots+\frac{1}{\sqrt{\{(2n-1)2n\}}}$$

prove that $f(n)$ tends to a limit l which satisfies the inequality $\frac{1}{4}\leqslant l\leqslant 1$.

[Show that $f(n)$ is a monotonic decreasing function of n.]

4. If $\phi(x)$ is continuous for $a\leqslant x\leqslant b$ and if $\lim_{x\to k}\{f(x)\} = l$, where k may be finite or infinite and $a\leqslant l\leqslant b$, prove that

$$\lim_{x\to k} \phi\{f(x)\} = \phi(l).$$

5. Evaluate

(i) $\overline{\lim_{x\to 0}}\left(\cos\dfrac{1}{x}\right)$, (ii) $\underline{\lim_{n\to\infty}}\{(-1)^n+\sin\frac{1}{4}n\pi\}$,

(iii) $\underline{\lim_{x\to 0}}\dfrac{1}{x+1}\sin\dfrac{1}{x}$, (iv) $\overline{\lim_{x\to 0}}\dfrac{1}{x+1}\sin\dfrac{1}{x}$.

6. If, as $x\to 0$, $f(x) = x+o(x^2)$, $\phi(x) = x^{-2}+O(x^{-1})$, prove that $f(x)\phi(x) = x^{-1}+O(1)$.

ANSWERS. 1. (i) $\frac{1}{2}$; (ii) a_0/b_0; (iii) $\frac{1}{2}$ if $|x|\leqslant 1$, x if $|x|>1$; (iv) x if $|x|<1$, 0 if $x = \pm 1$, $-1/x$ if $|x|>1$; (v) $\sqrt{2}$. 5. (i) 1; (ii) $-1-\frac{1}{2}\sqrt{2}$; (iii) -1; (iv) 1.

CHAPTER II

SOME PROPERTIES OF PARTICULAR FUNCTIONS

12. The Logarithmic and Exponential Functions. In this chapter we shall consider briefly some of the simpler functions of analysis. It will be assumed that the reader is familiar with their well-known properties, and we shall therefore confine ourselves to a discussion of those properties which are required for an adequate understanding of infinite series. We begin with the logarithmic and exponential functions.

The function $\log x$ is defined for $x>0$ by means of the integral $\int_1^x t^{-1}\,dt$. The relation $y = \int_1^x t^{-1}\,dt$, $x>0$, defines y as a monotonic increasing continuous function of x which tends to infinity as x tends to infinity, and it may be shown that, for all values of y, x is a positive, continuous, monotonic increasing function of y. This function we denote by $\exp y$. If the number e is defined by the equation

$$1 = \int_1^e t^{-1}\,dt,$$

it may be *proved* that, when y is a *rational* number (that is, a number of the form m/n where m and n are integers), $\exp y = e^y$. When y is irrational e^y is *defined* to be $\exp y$. The function a^y is then defined for all values of y and all positive values of a to be $e^{y \log a}$.

We now deduce some properties of these functions.

(i) *If x is any real number and n is any positive integer,*

$$e^x = 1 + \frac{x}{1!} + \frac{x^2}{2!} + \dots + \frac{x^{n-1}}{(n-1)!} + \frac{x^n}{n!}e^{\theta x},$$

where $0<\theta<1$.

Let $y = e^x$. Then $x = \log y = \int_1^y t^{-1}\,dt.$

Thus

$$\frac{dx}{dy} = y^{-1}, \quad \frac{dy}{dx} = y = e^x.$$

It follows at once that the derivatives of all orders of e^x are e^x and that their value when $x = 0$ is 1. The result then follows from Theorem C (ii).

An immediate consequence of this result is that, as $x \to 0$, $e^x = 1+x+O(x^2)$.

(ii) *If α and x are positive, there is a positive number K, independent of x, such that $e^x > Kx^\alpha$.*

Let $p = [\alpha]+1$. Since the terms on the right-hand side of the expansion for e^x in (i) are positive, we have by taking n large enough, for $x \geqslant 1$,

$$e^x > \frac{x^p}{p!} > \frac{x^\alpha}{p!},$$

and, for $0 \leqslant x < 1$,

$$e^x > \frac{x^{p-1}}{(p-1)!} \geqslant \frac{x^\alpha}{(p-1)!}.$$

The result follows by taking K to be $1/p!$.

(iii) *If α is any real number then, as $x \to \infty$,*

$$x^{-\alpha}e^x \to \infty, \quad x^\alpha e^{-x} \to 0.$$

Let β be any positive number greater than α. Then, by (ii).

$$x^{-\alpha}e^x > Kx^{-\alpha}x^\beta = Kx^{\beta-\alpha} \to \infty.$$

The second result follows at once from the first.

(iv) *If δ is any positive number then, as $x \to \infty$, $\log x = o(x^\delta)$ and, as $x \to 0$, $\log x = o(x^{-\delta})$.*

The second result follows from the first by writing $1/x$ for x. It is only necessary therefore to prove the first. Let $\log x = y/\delta$. Then we have to show that $ye^{-y}\to 0$ as $y\to\infty$, and this follows at once from (iii).

(v) *For $x\geqslant\rho>-1$ we have* *

$$\log(1+x) = x-\frac{x^2}{2}+\frac{x^3}{3}-\ldots+(-1)^{n-1}\frac{x^n}{n}+R(n, x),$$

where

$$|R(n, x)|<K\frac{|x|^{n+1}}{n+1},$$

K being positive and independent of n and x.

For $x>-1$ we have

$$\log(1+x)=\int_1^{1+x} t^{-1}\,dt = \int_0^x (1+t)^{-1}\,dt$$

$$=\int_0^x \{1-t+t^2-\ldots+(-1)^{n-1}t^{n-1}+(-1)^n t^n(1+t)^{-1}\}dt$$

$$=x-\frac{x^2}{2}+\frac{x^3}{3}-\ldots+(-1)^{n-1}\frac{x^n}{n}+R(n, x),$$

where

$$R(n, x) = (-1)^n\int_0^x t^n(1+t)^{-1}dt.$$

If † $x\geqslant 0$,

$$|R(n, x)|\leqslant\int_0^x t^n dt = \frac{x^{n+1}}{n+1},$$

while, if $-1<\rho\leqslant x<0$,

$$|R(n, x)| = \int_x^0 (-t)^n(1+t)^{-1}dt$$

$$\leqslant(1+\rho)^{-1}\int_x^0(-t)^n dt$$

$$= (1+\rho)^{-1}\frac{|x|^{n+1}}{n+1}.$$

The result therefore follows.

* In this inequality ρ is any fixed number greater than -1. If necessary it can be regarded as being as close to -1 as we please.

† See *G.*, p. 77.

The particular case $n = 2$ is of special importance. In this case we may write

$$\log(1+x) = x+O(x^2),\ (-1<\rho\leqslant x\leqslant 1),$$

or

$$\log(1+x)\sim x,\ (x\to 0).$$

The first of these is of course true for $x>1$, but is obviously a result of no mathematical significance.

(vi) *If α is any real number and $x>-1$, then*

$$(1+x)^\alpha = 1+\alpha x+\frac{\alpha(\alpha-1)}{1.2}x^2+\ldots+\frac{\alpha(\alpha-1)\ldots(\alpha-n+2)}{1.2\ldots(n-1)}x^{n-1}$$
$$+\frac{(1-\theta)^{n-1}}{(1+\theta x)^{n-\alpha}}\frac{\alpha(\alpha-1)\ldots(\alpha-n+1)}{(n-1)!}x^n,$$

where $0<\theta<1$.

This follows almost at once from the second form of Theorem C (ii) since, when α is any real number and $x>-1$,

$$\frac{d}{dx}(1+x)^\alpha = \frac{d}{dx}\{e^{\alpha\log(1+x)}\} = \frac{\alpha}{1+x}e^{\alpha\log(1+x)} = \alpha(1+x)^{\alpha-1},$$

which is equal to α when $x = 0$. The values of the other derivatives at $x = 0$ then follow without difficulty.

It follows from this result that, for $-1<x<1$,

$$(1+x)^\alpha = 1+\alpha x+O(x^2),$$
$$(1+x)^\alpha = 1+\alpha x+\frac{\alpha(\alpha-1)}{1.2}x^2+O(|x|^3), \text{ etc.}$$

(vii) *If x is any real number, then*

$$\lim_{y\to\pm\infty}\left(1+\frac{x}{y}\right)^y = e^x.$$

Write $h = x/y$ so that, as $y\to\pm\infty$, $h\to 0$. Then, since we may assume that $h>-1$,

$$\left(1+\frac{x}{y}\right)^y = e^{\frac{x}{h}\log(1+h)}$$
$$= e^{x\{1+O(|h|)\}}$$
$$\to e^x,$$

as $h\to 0$, since the exponential function is continuous at the origin.

(viii) *If β is any real number, then*

$$\lim_{x\to a}\frac{x^\beta-a^\beta}{x-a}=\beta a^{\beta-1}.$$

Write $\dfrac{x}{a}=1+h$. Then supposing, as we may, that $h>-1$,

$$\begin{aligned}\lim_{x\to a}\frac{x^\beta-a^\beta}{x-a}&=\lim_{h\to 0}\frac{(1+h)^\beta-1}{h}\,.\,a^{\beta-1}\\&=a^{\beta-1}\lim_{h\to 0}\frac{e^{\beta\log(1+h)}-1}{\beta\log(1+h)}\cdot\frac{\beta\log(1+h)}{h}\\&=\beta a^{\beta-1}\lim_{h\to 0}\frac{e^{\beta\log(1+h)}-1}{\beta\log(1+h)}\\&=\beta a^{\beta-1},\end{aligned}$$

by (i) since $\beta\log(1+h)\to 0$ as $h\to 0$.

13. The Hyperbolic Functions. The hyperbolic functions are defined by the relations

$$\sinh x=\tfrac{1}{2}(e^x-e^{-x}),\ \cosh x=\tfrac{1}{2}(e^x+e^{-x}),$$
$$\tanh x=\frac{\sinh x}{\cosh x},\ \coth x=\frac{1}{\tanh x},$$
$$\operatorname{sech} x=\frac{1}{\cosh x},\quad \operatorname{cosech} x=\frac{1}{\sinh x}.$$

These relations define $\sinh x$, $\cosh x$, $\tanh x$, $\operatorname{sech} x$ for all values of x, and $\coth x$, $\operatorname{cosech} x$ for all values of x except zero.

The following properties of these functions may be easily verified :—

(i) $\cosh^2 x-\sinh^2 x=1$,

(ii) $\sinh(x+y)=\sinh x\cosh y+\cosh x\sinh y$,
$\cosh(x+y)=\cosh x\cosh y+\sinh x\sinh y$,

(iii) $\dfrac{d}{dx}\sinh x=\cosh x$, $\dfrac{d}{dx}\cosh x=\sinh x$,

(iv) $$\sinh x = x+\frac{x^3}{3!}+\frac{x^5}{5!}+\ldots+\frac{x^{2n-1}}{(2n-1)!}+\frac{x^{2n}}{(2n)!}\sinh\theta_1 x,$$

$$\cosh x = 1+\frac{x^2}{2!}+\frac{x^4}{4!}+\ldots+\frac{x^{2n}}{(2n)!}+\frac{x^{2n+1}}{(2n+1)!}\sinh\theta_2 x,$$

where $0<\theta_1<1$, $0<\theta_2<1$.

14. The Circular Functions. It is not necessary to discuss here logical definitions of the circular functions. We assume that the reader is familiar with the properties of these functions, and we confine ourselves to a statement of the Taylor expansions of $\sin x$ and $\cos x$. We have

$$\sin x = x-\frac{x^3}{3!}+\frac{x^5}{5!}-\ldots+(-1)^{n-1}\frac{x^{2n-1}}{(2n-1)!}+\frac{x^{2n}}{(2n)!}\sin(\theta_1 x+n\pi),$$

$$\cos x=1-\frac{x^2}{2!}+\frac{x^4}{4!}-\ldots+(-1)^n\frac{x^{2n}}{(2n)!}+\frac{x^{2n+1}}{(2n+1)!}\sin\{\theta_2 x+(n+1)\pi\},$$

where $0<\theta_1<1$, $0<\theta_2<1$.

Examples

1. Evaluate the limits

(i) $\lim\limits_{n\to\infty}\left(1+\frac{1}{n}\right)^a$, (ii) $\lim\limits_{n\to\infty}\left(1+\frac{1}{n}\right)^{n^a}$,

(iii) $\lim\limits_{x\to\infty} x\sin\frac{1}{x}$, (iv) $\lim\limits_{n\to\infty}\frac{(n+1)\log n-n\log(n+1)}{\log n}$.

2. Evaluate the limits

(i) $\lim\limits_{x\to 0} x^x$, (ii) $\lim\limits_{x\to 0} x^{x^x}$, (iii) $\lim\limits_{x\to 0}(\sin x)^{\tan x}$.

3. Prove that, as $x\to\infty$,

$$\frac{\log\log x}{\log x}\to 0, \quad \frac{x\log\log x}{(\log x)^n}\to\infty.$$

4. Prove that, as $x \to 0$,

(i) $e^{2x}-2e^{x}+1 \sim x^2$,

(ii) $e^{x^2} \sin x = x+O(|x|^3)$,

(iii) $e^{x} \cos x \log (1+x) = x+O(x^2)$,

(iv) $\dfrac{x^2+\log (1-x^2)}{x^2-\sinh^2 x} \to 3/2$,

and that, as $x \to \infty$,

(v) $\log (1+e^{x}+e^{x^2}) \sim x^2$.

5. Prove that as $x \to 0$, $\sinh x \sim x$, and that as $x \to \infty$, $\log \sinh x \sim x$.

6. Prove that,

$$e^{x} \sin x = x+\frac{2x^2}{2!}+\frac{2x^3}{3!}-\frac{2^2 x^5}{5!} \cdots$$

$$+\frac{2^{\frac{1}{2}(n-1)} \sin \left(\dfrac{n-1}{4}\pi\right) x^{n-1}}{(n-1)!}+\frac{2^{\frac{1}{2}n} x^n}{n!} e^{\theta x} \sin (\theta x+\tfrac{1}{4} n\pi),$$

where $0<\theta<1$.

ANSWERS. 1. (i) 1; (ii) ∞ if $a>1$, e if $a=1$, 1 if $a<1$; (iii) 1; (iv) 1. 2. (i) 1; (ii) 0; (iii) 1.

CHAPTER III

REAL SEQUENCES AND SERIES

15. Definition of a Sequence. Suppose that A_n is a function of the positive integral variable n which is defined for all values of n. Then the ordered set of numbers

$$A_1, A_2, A_3, \ldots, A_n, \ldots$$

obtained from A_n by giving n the values 1, 2, ... in turn is called an **infinite sequence** or, more simply, a **sequence.** The numbers A_1, A_2, ... are called respectively the first, second, ... terms of the sequence. In this chapter we shall assume that our sequences are real, that is, have only real terms.

16. Convergent, Divergent and Oscillating Sequences. The sequence A_1, A_2, ... is said to **converge** or **to be convergent** to the sum a if* $\lim A_n = a$. If A_n does not tend to a finite limit the sequence is said to be **divergent.** Divergent sequences are often classified further into sequences which are **properly divergent,** or **oscillate finitely** or **oscillate infinitely.** In the first of these $\lim A_n = \pm\infty$, in the second A_n is a bounded function of n and in the third A_n is not bounded. For example, the sequences for which A_n is equal to $1+1/n$, $\log n$, $\sin\frac{1}{6}n\pi$, $(-1)^n n$, are respectively convergent, properly divergent, finitely oscillating and infinitely oscillating.

* By $\lim A_n = a$ we mean $\lim\limits_{n\to\infty} A_n = a$. We shall adopt this contracted notation throughout when dealing with functions of the positive integral variable n.

17. Infinite Series. Suppose that a_n is a function of the positive integral variable n. Let

$$A_n = a_1 + a_2 + \ldots + a_n = \sum_{r=1}^{n} a_r.$$

The function A_n is called the sum to n terms or the n-th **partial sum** of the series $a_1+a_2+a_3+\ldots$. This series is often denoted by Σa_n, or, more precisely, by $\sum_{n=1}^{\infty} a_n$, and $a_1, a_2, \ldots$ are called respectively the first, second, ... terms of the series. The series $a_1+a_2+a_3+\ldots$ is said to converge, properly diverge, oscillate finitely or oscillate infinitely according as the sequence $A_1, A_2, A_3, \ldots$ converges, properly diverges, oscillates finitely or oscillates infinitely. If $\lim A_n = a$, where a is finite, the series Σa_n is said to converge to the sum a, and we write $\sum_{n=1}^{\infty} a_n = a$.

It will be observed that, in the above paragraph, the notation $\sum_{n=1}^{\infty} a_n$ has been employed in two different senses. It was used firstly as a means of naming a particular series and secondly as the sum of the series. The reader will find that no difficulty arises from this ambiguity in notation.

18. Important Particular Series. We now obtain the expansions in infinite series of certain well-known functions.

(i) *If* $-1 < x \leqslant 1$, *we have*

$$\log(1+x) = x - \frac{x^2}{2} + \frac{x^3}{3} - \ldots = \sum_{n=1}^{\infty} (-1)^{n-1} \frac{x^n}{n}.$$

Let A_n denote the n-th partial sum of the series on the right. Then, from Arts. **12** and **5**, we have, for $-1 < \rho \leqslant x \leqslant 1$,

$$|A_n - \log(1+x)| < K \frac{|x|^{n+1}}{n+1} \to 0$$

as $n \to \infty$. That is, the series on the right is convergent and has the sum $\log(1+x)$ when $-1 < x \leqslant 1$.

When $x = 1$ we obtain the following interesting result:

$$\log 2 = 1 - \tfrac{1}{2} + \tfrac{1}{3} - \tfrac{1}{4} + \ldots .$$

(ii) *For all values of x we have*

$$e^x = 1 + \frac{x}{1!} + \frac{x^2}{2!} + \ldots = \sum_{n=0}^{\infty} \frac{x^n}{n!}.$$

For the series on the right we have, from Arts. **12** and **5**,

$$|A_n - e^x| = e^{\theta x} \frac{|x|^n}{n!} \to 0,$$

as $n \to \infty$ for all values of x. The result follows.

(iii) *For all values of x we have*

$$\sin x = x - \frac{x^3}{3!} + \frac{x^5}{5!} - \ldots,$$

$$\cos x = 1 - \frac{x^2}{2!} + \frac{x^4}{4!} - \ldots .$$

(iv) *For all values of x we have*

$$\sinh x = x + \frac{x^3}{3!} + \frac{x^5}{5!} + \ldots,$$

$$\cosh x = 1 + \frac{x^2}{2!} + \frac{x^4}{4!} + \ldots .$$

The proofs of (iii) and (iv) are similar to the proof of (ii).

(v) *If $-1 < x < 1$ and a is any real number we have*

$$(1+x)^a = 1 + ax + \frac{a(a-1)}{1.2} x^2 + \ldots = \sum_{n=0}^{\infty} \frac{a(a-1)\ldots(a-n+1)}{1.2\ldots n} x^n.$$

When a is a positive integer the series on the right reduces to a finite sum and the expansion is then valid for all values of x.

We may confine our attention to the first part of the theorem, the second part being merely a statement of the "positive integral index case" of the binomial theorem.

From Art. **12** we have, for the series on the right-hand side, when $x>-1$,

$$|A_n-(1+x)^a| = (1+\theta x)^{a-1}\left(\frac{1-\theta}{1+\theta x}\right)^{n+1} K_n|x|^n,$$

where

$$K_n = \left|\frac{(n-1-a)(n-2-a)\ldots(-a)}{(n-1)(n-2)\ldots 2.1}\right|.$$

Suppose that a is positive. Let $p=[a]$. Then, for all sufficiently large values of n,

$$\begin{aligned} K_n &= \left\{\frac{(n-1-a)(n-2-a)\ldots(p+1-a)}{(n-1)(n-2)\ldots(p+1)}\right\} \\ &\qquad\qquad \times\left\{\frac{(a-p)(a+1-p)\ldots(a-1)a}{p(p-1)\ldots 2.1}\right\} \\ &= \left(1-\frac{a}{n-1}\right)\left(1-\frac{a}{n-2}\right)\cdots\left(1-\frac{a}{p+1}\right).O(1) \\ &= O(1), \end{aligned}$$

since each factor lies between 0 and 1.

Suppose now that a is negative. Let $\beta=-a$, $q=[\beta]$. Then, for all sufficiently large values of n,

$$\begin{aligned} K_n &= \left\{\frac{(n+\beta-1)(n+\beta-2)\ldots(n+\beta-q-1)}{1.2\ldots q}\right\} \\ &\qquad\qquad \times\left\{\frac{(n+\beta-q-2)\ldots(\beta+1)\beta}{(n-1)(n-2)\ldots(q+1)}\right\} \\ &< \frac{(n+\beta)^{q+1}}{q!}\left(1-\frac{q+1-\beta}{n-1}\right)\left(1-\frac{q+1-\beta}{n-2}\right)\cdots\left(1-\frac{q+1-\beta}{q+1}\right) \\ &< \frac{(2n)^{q+1}}{q!} \\ &= O(n^{q+1}). \end{aligned}$$

It follows from Art. **5** that, whether a be positive or negative, $|K_n|x^n\to 0$ when $-1<x<1$.

Moreover, when $x>-1$,

$$0<(1-\theta)/(1+\theta x)<(1-\theta)/(1-\theta)=1.$$

The result stated then follows at once.

Two important particular cases of this result are worth stating independently. Putting $a = -1$ and $-x$ for x we obtain

$$(1-x)^{-1} = 1+x+x^2+\ldots, \ (-1<x<1).$$

Again, putting $a = -2$ and $-x$ for x, we obtain

$$(1-x)^{-2} = 1+2x+3x^2+\ldots, \ (-1<x<1).$$

19. The General Principle of Convergence. We now give a very general criterion for the convergence of an infinite series or sequence.

THEOREM 6. *A necessary and sufficient condition for the sequence $A_1, A_2, \ldots$ to be convergent is that, given ϵ, there should exist a positive integer $N = N(\epsilon)$ such that $|A_{n+p}-A_n|<\epsilon$ for all integral values of $n>N$ and for all positive integral values of p.*

If the sequence is convergent there is a finite number a such that $A_n \to a$. Hence, given ϵ, we can find a positive integer $N = N(\epsilon)$ such that $|A_n-a|<\frac{1}{2}\epsilon$ whenever $n>N$. Thus, if p is any positive integer and $n>N$,

$$|A_{n+p}-A_n| \leqslant |A_{n+p}-a|+|A_n-a|<\tfrac{1}{2}\epsilon+\tfrac{1}{2}\epsilon = \epsilon.$$

Thus the condition is necessary.

On the other hand, if the condition is satisfied, it follows that $|A_n-A_{N+1}|<\epsilon$ for all values of $n \geqslant N+1$; that is, for $n \geqslant N+1$,

$$A_{N+1}-\epsilon<A_n<A_{N+1}+\epsilon,$$

so that A_n is bounded. It follows that both $\overline{\lim}\, A_n$ and $\underline{\lim}\, A_n$ are finite and that *

$$A_{N+1}-\epsilon \leqslant \underline{\lim}\, A_n \leqslant \overline{\lim}\, A_n \leqslant A_{N+1}+\epsilon,$$

whence

$$0 \leqslant \overline{\lim}\, A_n - \underline{\lim}\, A_n \leqslant 2\epsilon.$$

But ϵ is arbitrary, so that $\overline{\lim}\, A_n = \underline{\lim}\, A_n$. Hence $\lim A_n$ exists and is finite. The sequence is therefore convergent.

* Of the three equality or inequality signs in the succeeding line one at least must not be the equality sign.

The following is the analogue of Theorem 6 for series.

THEOREM 7. *A necessary and sufficient condition for the series Σa_n to be convergent is that, given ϵ, we can find $N = N(\epsilon)$, such that*

$$\left| \sum_{\nu=n+1}^{n+p} a_\nu \right| < \epsilon,$$

for all integral values of $n > N$ and all positive integral values of p.

This follows at once from Theorem 6, for, if

$$A_n = \sum_{\nu=1}^{n} a_\nu,$$

we have

$$A_{n+p} - A_n = a_{n+1} + a_{n+2} + \ldots + a_{n+p} = \sum_{\nu=n+1}^{n+p} a_\nu.$$

The following deduction from Theorem 7 is very important.

THEOREM 8. *The series Σa_n cannot be convergent unless $a_n \to 0$.*

The theorem will be proved if we show that, if Σa_n is convergent, $a_n \to 0$. This follows from Theorem 7, with $p = 1$, or from the definition of convergence since A_n, A_{n-1} both tend to the same limit and $a_n = A_n - A_{n-1}$.

It is important to notice that the condition $a_n \to 0$ does not necessarily imply the convergence of the series Σa_n. For example, in the case of the harmonic series $1 + \frac{1}{2} + \frac{1}{3} + \frac{1}{4} + \ldots$, we have $a_n = n^{-1} \to 0$ and

$$A_{2n} - A_n = \frac{1}{n+1} + \frac{1}{n+2} + \ldots + \frac{1}{2n} > \frac{n}{2n} = \tfrac{1}{2}.$$

This inequality holds for all values of n however large, so that by Theorem 7 the series is not convergent.

Example.—Show that the series $\Sigma \sin(n\theta + \phi)$, $\Sigma \cos(n\theta + \phi)$, where ϕ is any real number and θ is any real number except zero or a multiple of 2π, oscillate finitely.

When $\theta = \pi$ the series reduce to

$$\sin\phi\ \Sigma(-1)^n,\ \cos\phi\ \Sigma(-1)^n$$

which oscillate finitely. Hence we may confine ourselves to values of θ for which $0<\theta<\pi$ and for which $\pi<\theta<2\pi$. We shall show first that, for such values of θ, $\sin(n\theta+\phi)$ does not tend to zero as $n\to\infty$.

Let $\kappa_n = \sin(n\theta+\phi)$ and suppose that $\kappa_n\to 0$. Then we can find N such that, whenever $n>N$,

$$|\kappa_n| < \frac{|\sin\theta|}{\sqrt{(4+\sin^2\theta)}};$$

that is,

$$|\kappa_{n+1}|+|\kappa_n| < \frac{2\,|\sin\theta|}{\sqrt{(4+\sin^2\theta)}} \quad . \quad . \quad . \quad . \quad (1)$$

Now

$$\begin{aligned}|\kappa_{n+1}| &= |\kappa_n\cos\theta+\cos(n\theta+\phi)\sin\theta| \\ &\geqslant \sqrt{(1-\kappa_n^2)}\,|\sin\theta|-|\kappa_n|\,|\cos\theta|,\end{aligned}$$

so that

$$\begin{aligned}|\kappa_{n+1}|+|\kappa_n| &\geqslant \sqrt{(1-\kappa_n^2)}\,|\sin\theta| \\ &> \sqrt{\left(1-\frac{\sin^2\theta}{4+\sin^2\theta}\right)}\,|\sin\theta| = \frac{2\,|\sin\theta|}{\sqrt{(4+\sin^2\theta)}},\end{aligned}$$

which contradicts (1). Hence $\sin(n\theta+\phi)$ does not tend to zero. Since $\cos(n\theta+\phi) = \sin(n\theta+\phi+\frac{1}{2}\pi)$ it follows that $\cos(n\theta+\phi)$ does not tend to zero. By Theorem 8 it follows that the series $\Sigma\sin(n\theta+\phi)$, $\Sigma\cos(n\theta+\phi)$ are not convergent. They oscillate finitely, since

$$\begin{aligned}\left|\sum_{\nu=1}^{n}\begin{matrix}\sin\\ \cos\end{matrix}(\nu\theta+\phi)\right| &= \left|\begin{matrix}\sin\\ \cos\end{matrix}\{\tfrac{1}{2}(n+1)\theta+\phi\}\sin\tfrac{1}{2}n\theta\operatorname{cosec}\tfrac{1}{2}\theta\right| \\ &\leqslant |\operatorname{cosec}\tfrac{1}{2}\theta|.\end{aligned}$$

The reader will find it an interesting exercise to deduce from the fact that $\sin(n\theta+\phi)$, $\cos(n\theta+\phi)$ do not tend to zero, that these functions do not tend to a limit at all.

20. Some Preliminary Theorems on Series. We now show that infinite series possess certain of the well-known properties of finite sums.

THEOREM 9. *If* $\sum_{n=1}^{\infty} a_n = a$ *then* $\sum_{n=1}^{\infty} ca_n = ca$, *where* c *is any number independent of* n.

This follows at once from the identity

$$\sum_{r=1}^{n} ca_r = c\sum_{r=1}^{n} a_r$$

on making n tend to infinity.

THEOREM 10.

If $\sum_{n=1}^{\infty} a_n = a$, $\sum_{n=1}^{\infty} b_n = \beta$ *then* $\sum_{n=1}^{\infty}(a_n+b_n) = a+\beta$.

This follows from the identity

$$\sum_{r=1}^{n}(a_r+b_r) = \sum_{r=1}^{n} a_r+\sum_{r=1}^{n} b_r$$

by making n tend to infinity.

The same proof shows that if one of the given series Σa_n, Σb_n is divergent and if the other is convergent then the series $\Sigma(a_n+b_n)$ is also divergent.

THEOREM 11.

If $\sum_{n=1}^{\infty} a_n = a$ *then* $\sum_{n=0}^{\infty} a_n = a+a_0$ *and* $\sum_{n=2}^{\infty} a_n = a-a_1$.

We shall prove only the first part of the theorem, the proof of the second part being similar.

Let $A'_n = \sum_{\nu=0}^{n} a_\nu$. Then clearly $A'_n = a_0+A_n$. The result then follows on letting n tend to infinity.

It is clear that, if the series $\sum_{n=1}^{\infty} a_n$ is divergent, each of the series $\sum_{n=0}^{\infty} a_n$, $\sum_{n=2}^{\infty} a_n$ is also divergent.

The theorem shows that a new term may be inserted at the beginning of a convergent series without affecting its convergence, and that the first term may be removed from a convergent series without affecting its convergence. A trivial modification of the argument shows that the term inserted or removed need not necessarily be at the beginning of the series. A further slight extension enables us to conclude that the insertion or removal of any finite number of terms from a convergent series does not affect its convergence, and that the sums of the various series are related in the expected way.

THEOREM 12. *If the series Σa_n converges to the sum a then so does any series obtained from Σa_n by grouping the terms in brackets without altering the order of the terms.*

Suppose that the series derived from Σa_n by the insertion of brackets is Σb_ν and let B_ν denote the ν-th partial sum of the series Σb_ν. Suppose that B_ν contains n_ν terms of the given series. Then, since the order of the terms is unaltered, $B_\nu = A_{n_\nu}$. As $\nu \to \infty$, $n_\nu \to \infty$ and $A_{n_\nu} \to a$. It follows that $B_\nu \to a$, and the theorem is proved.

A similar result clearly holds for series which are properly divergent.

It should be noted that the converse of this theorem is false. For example, the series $(1-1)+(1-1)+\ldots$ is convergent, whereas the series obtained by removing brackets is not. Brackets may thus be inserted without affecting convergence but may *not* be removed.

Examples

1. By finding their n-th partial sums examine the convergence of the series :—

(i) $\sum\limits_{n=1}^{\infty} x^n$, (ii) $\sum\limits_{n=1}^{\infty} (an+b)x^n$, (iii) $\sum\limits_{n=1}^{\infty} \dfrac{1}{n(n+1)}$,

(iv) $\sum\limits_{n=1}^{\infty} \dfrac{2}{n(n+1)(n+2)}$, (v) $\sum\limits_{n=1}^{\infty} \dfrac{1}{(1+nx)\{1+(n+1)x\}}$,

(vi) $\sum\limits_{n=1}^{\infty} \dfrac{n}{(n+1)!}$.

2. Prove that, for $-\frac{1}{2}<x<\frac{1}{2}$,

$$\frac{9x}{(1-x)^2(1+2x)} = \sum_{n=1}^{\infty} \{3n+2+(-1)^{n+1}\, 2^{n+1}\}x^n.$$

3. Prove that

(i) $\log\{(1+x)^{1+x}\}+\log\{(1-x)^{1-x}\}$

$$=x^2+\frac{x^4}{2.3}+\frac{x^6}{3.5}+\frac{x^8}{4.7}+\ldots,\ (-1<x<1),$$

(ii) $2\log x-\log(x+1)-\log(x-1)$

$$=\frac{1}{x^2}+\frac{1}{2x^4}+\frac{1}{3x^6}+\ldots,\ (|x|>1),$$

(iii) $(1+x)e^{-x}-(1-x)e^{x}=\sum_{n=1}^{\infty}\frac{4n}{(2n+1)!}x^{2n+1},$

(iv) $\frac{1}{2}\log x=\frac{x-1}{x+1}+\frac{1}{3}\left(\frac{x-1}{x+1}\right)^3+\frac{1}{5}\left(\frac{x-1}{x+1}\right)^5+\ldots,\ (x>0).$

4. If $\sum_{n=1}^{\infty}a_n=a$, prove that $\sum_{n=1}^{\infty}(a_n+a_{n+1})=2a-a_1$.

5. If Σa_n is convergent show that the series $\Sigma\frac{n+1}{n}a_n$ is also convergent.

6. Prove that the series

$$\sum_{n=2}^{\infty}\frac{1}{n\log n},\quad \sum_{n=1}^{\infty}n^{-a},\ (a\leqslant 1),$$

are properly divergent.

7. Show that the series

$$\sum_{n=1}^{\infty}\log\left(1+\frac{1}{n}\right),\quad \sum_{n=1}^{\infty}\sin\tfrac{1}{4}n\pi,\quad \sum_{n=1}^{\infty}n\sin\tfrac{1}{4}n\pi,$$

are respectively properly divergent, finitely oscillating, infinitely oscillating.

8. Prove that the series $\sum_{n=1}^{\infty}\frac{1}{n(2n-1)}$ converges to the sum $2\log 2$, and that the series $\sum_{n=1}^{\infty}(-1)^{n-1}\frac{2n+1}{n(n+1)}$ converges to the sum 1.

9. If $\sum_{n=1}^{\infty}a_n$ oscillates finitely and if $a_n=o\left(\frac{1}{n}\right)$ show that $\sum_{n=2}^{\infty}n(a_n-a_{n-1})$ also oscillates finitely.

10. Using the relation

$$\tan \tfrac{1}{2}x = \cot \tfrac{1}{2}x - 2 \cot x,$$

find the sum of the series $\sum_{n=1}^{\infty} \frac{1}{2^n} \tan \frac{x}{2^n}$.

11. A sequence of positive terms $A_1, A_2, \ldots, A_n, \ldots$ satisfies the relation

$$A_{n+1} = \frac{3(1+A_n)}{3+A_n}.$$

Show that A_n is a monotonic decreasing or increasing function of n according as $A_1 \gtrless \sqrt{3}$. Deduce the value of $\lim A_n$.

12. If $x_1 = \cos \theta$, $y_1 = 1$ and

$$x_{n+1} = \tfrac{1}{2}(x_n + y_n),\ y_{n+1} = \sqrt{(x_{n+1} y_n)},\ n = 1, 2, \ldots,$$

show that x_n and y_n tend to the common limit $\sin \theta / \theta$.

13. If, for all values of n, $b_n \geqslant 0$, if $\lim (b_1 + b_2 + \ldots + b_n) = \infty$, and if $A_n \to a$, prove that

$$\lim \frac{b_1 A_1 + b_2 A_2 + \ldots + b_n A_n}{b_1 + b_2 + \ldots + b_n} = a.$$

Deduce that

(i) $$\frac{\sin \theta + \sin \frac{\theta}{2} + \ldots + \sin \frac{\theta}{n}}{1 + \frac{1}{2} + \frac{1}{3} + \ldots + \frac{1}{n}} \to \theta,$$

(ii) $$\frac{1}{n^2} \left\{ 1^2 \sin \theta + 2^2 \sin \frac{\theta}{2} + \ldots + n^2 \sin \frac{\theta}{n} \right\} \to \tfrac{1}{2}\theta.$$

ANSWERS. 1. (i) Convergent if $|x| < 1$, properly divergent if $x \geqslant 1$, oscillates between -1 and 0 if $x = -1$, oscillates infinitely if $x < -1$; (ii) convergent if $|x| < 1$, properly divergent if $x \geqslant 1$, oscillates infinitely if $x \leqslant -1$; (iii) convergent; (iv) convergent; (v) convergent if $x \neq 0$, properly divergent if $x = 0$, [If $x = -1/n$, where n is a positive integer, the series is meaningless.]; (vi) convergent. 10. $1/x - \cot x$, where $x \neq 0$. 11. $\sqrt{3}$.

CHAPTER IV

SERIES OF NON-NEGATIVE TERMS

21. A Fundamental Theorem. We now consider in some detail series whose non-zero terms are all of the same sign. We shall assume that we are dealing with a series Σa_n where $a_n \geqslant 0$ for all values of n. There is no loss of generality in so doing, for a series Σa_n for which $a_n \leqslant 0$ falls into this category when we multiply by -1. It is almost intuitive to expect that such a series cannot oscillate. The theorem below contains a formal statement and proof of this result.

THEOREM 13. *If $a_n \geqslant 0$ the series Σa_n is either convergent or properly divergent.*

Since $a_n \geqslant 0$, A_n is a monotonic increasing function of n. The result then follows from Theorem 3.

22. Rearrangement of Terms. We have already seen in Art. **20** that the terms of *any* convergent series may be grouped in brackets without destroying its convergence so long as the order of the terms is not altered. For series of non-negative terms we shall show that we may dispense with the latter condition and we shall also show that brackets may be *removed* as well as inserted without affecting convergence.

THEOREM 14. *Suppose that $a_n \geqslant 0$. Let Σb_n be any series whose terms are those of the series Σa_n in a different order. If the series Σa_n converges to a then so does Σb_n, and if Σa_n is properly divergent then so is Σb_n.*

Suppose that

$$b_1 = a_{m_1}, \quad b_2 = a_{m_2}, \quad b_3 = a_{m_3}, \ldots$$

Then, if p is the largest of the integers $m_1, m_2, \ldots, m_n$ and * $B_n = \sum_{\nu=1}^{n} b_\nu$, we have

(i) $$B_n = \sum_{\nu=1}^{n} a_{m_\nu} \leqslant \sum_{\nu=1}^{p} a_\nu = A_p$$

where $p \to \infty$ as $n \to \infty$.

A similar argument shows that there is an integer k which tends to infinity as $n \to \infty$ such that

(ii) $$A_n \leqslant B_k.$$

Suppose now that $\sum_{n=1} a_n = a$. Then (i) shows that $\lim B_n \leqslant a$, while (ii) shows that $\lim B_k \geqslant a$. It therefore follows that Σb_n converges to a.

If, on the other hand, Σa_n is properly divergent then (ii) shows that $B_k \to \infty$, so that Σb_n is also properly divergent.

THEOREM 15. *Suppose that $a_n \geqslant 0$ and that Σb_n is a series obtained from Σa_n by picking terms at random and grouping in brackets in any way. If the series Σb_n converges to a then so does Σa_n and if the series Σb_n is properly divergent then so is Σa_n.*

Let Σc_n be the series Σb_n with the brackets removed and the order of the terms unaltered. Suppose first that Σb_n converges to the sum a. Then Σc_n must converge to the sum a for, if it were to converge to a sum other than a or were properly divergent it would follow from Theorem 12 that Σb_n could not converge to the sum a. From Theorem 14 it then follows that Σa_n converges to the sum a.

If Σb_n is properly divergent the same type of argument shows at once that Σa_n must also be properly divergent.

These two theorems show in effect that, as regards the alteration of the order of terms and the insertion of brackets, series of non-negative terms behave exactly like finite sums.

* The notation $B_n = \sum_{\nu=1}^{n} b_\nu$ will be adhered to throughout the book.

23. Tests for Convergence. When A_n can be calculated explicitly it is usually easy to determine whether or not the series Σa_n is convergent. For a very large number of interesting series, however, it is not possible to calculate A_n. It is therefore of some importance to obtain tests for the convergence of series which involve only simple properties of the terms themselves. A test for general series has already been obtained in Theorem 6, but this test is not of the type which we are seeking, for, naturally enough, to evaluate or obtain inequalities involving the expression $\sum_{\nu=n+1}^{n+p} a_\nu$ is hardly less awkward as a rule than the evaluation of A_n.

The tests which follow, although stated for series whose terms are *all* non-negative, hold also for series whose terms are non-negative only from some value of n onwards.

24. The Integral Test. This test is applicable only in the case of series Σa_n for which a_n is a monotonic decreasing function. We prove first an important auxiliary theorem:

THEOREM 16. *If, for* $x \geqslant 1$, $f(x)$ *is a non-negative, monotonic decreasing integrable function such that* $f(n) = a_n$ *for all positive integral values of* n, *then*

$$\lim \left\{A_n - \int_1^n f(x)dx\right\}$$

exists and satisfies the inequality

$$0 \leqslant \lim \left\{A_n - \int_1^n f(x)dx\right\} \leqslant a_1.$$

By hypothesis we have,* for all positive integral values of r,

$$\int_r^{r+1} f(r)dx \geqslant \int_r^{r+1} f(x)dx \geqslant \int_r^{r+1} f(r+1)dx\,;$$

* See *G.*, p. 77.

that is,

$$a_r \geqslant \int_r^{r+1} f(x)dx \geqslant a_{r+1}.$$

Give r the values 1, 2, ..., $n-1$ in succession and add. Then

$$A_{n-1} \geqslant \int_1^n f(x)dx \geqslant A_n - a_1;$$

that is

$$a_n - A_n \leqslant -\int_1^n f(x)dx \leqslant a_1 - A_n$$

or

$$a_n \leqslant A_n - \int_1^n f(x)dx \leqslant a_1$$

and, *a fortiori*,

$$0 \leqslant A_n - \int_1^n f(x)dx \leqslant a_1.$$

Now $A_n - \int_1^n f(x)dx$ is a monotonic decreasing function of n, for

$$\left\{A_n - \int_1^n f(x)dx\right\} - \left\{A_{n+1} - \int_1^{n+1} f(x)dx\right\}$$
$$= \int_n^{n+1} f(x)dx - a_{n+1} \geqslant 0.$$

It follows from Theorem 3 that $A_n - \int_1^n f(x)dx$ tends to a limit which satisfies the inequality stated.

THEOREM 17. *If, for $x \geqslant 1$, $f(x)$ is a non-negative, monotonic decreasing integrable function such that $f(n) = a_n$ for all positive integral values of n, then the series $\sum_{n=1}^{\infty} a_n$ and the integral $\int_1^{\infty} f(x)dx$ converge or diverge together. In other* *words, if one of them converges so does the other and if one of them properly diverges so does the other.*

These results follow at once from Theorem 16, since

$$A_n = \int_1^n f(x)dx + \left\{A_n - \int_1^n f(x)dx\right\}.$$

Suppose that $f(x) = x^{-\lambda}$. If $\lambda \geqslant 0$, $f(x)$ satisfies the conditions of Theorem 17 for $x>0$. Also $\int_1^\infty x^{-\lambda}dx$ is convergent if $\lambda>1$ and properly divergent if $0 \leqslant \lambda \leqslant 1$. Thus the series

$$1 + \frac{1}{2^\lambda} + \frac{1}{3^\lambda} + \ldots$$

is convergent for $\lambda>1$ and properly divergent for $0 \leqslant \lambda \leqslant 1$. It is also properly divergent for $\lambda<0$ since in this case its n-th term does not tend to zero.

Suppose now that $f(x) = x^{-1}(\log x)^{-\lambda}$. Then, for $\lambda \geqslant 0$ and $x>1$, $f(x)$ satisfies the conditions of Theorem 17. Moreover,

$$\int_2^\infty \frac{dx}{x(\log x)^\lambda} = \int_{\log 2}^\infty \frac{du}{u^\lambda}$$

which is convergent if $\lambda>1$ and properly divergent for $0 \leqslant \lambda \leqslant 1$.

Similar arguments show that the set of series

$$\sum_{n=3}^{\infty} \frac{1}{n \log n\, (\log \log n)^\lambda}, \sum_{n=16}^{\infty} \frac{1}{n \log n \log \log n\, (\log \log \log n)^\lambda}, \ldots$$

are convergent for $\lambda>1$ and properly divergent for $0 \leqslant \lambda \leqslant 1$.

25. The Comparison Tests.

THEOREM 18. *If $a_n \geqslant 0$, $b_n \geqslant 0$, if there is a positive number K, independent of n, and an integer N such that $a_n < Kb_n$ whenever $n>N$ and if Σb_n is convergent, then Σa_n is also convergent.*

Since Σb_n is convergent, given ϵ, we can find $N_1 = N_1(\epsilon)$ such that, for all values of $n > N_1$ and all positive integral values of p,

$$\sum_{\nu=n+1}^{n+p} b_\nu < \epsilon/K.$$

Let $N_2 = \text{Max}\ (N, N_1)$. Then, whenever $n > N_2$, we have

$$\sum_{\nu=n+1}^{n+p} a_\nu < K \sum_{\nu=n+1}^{n+p} b_\nu < \epsilon,$$

and, since this inequality holds for all positive integral values of p, it follows that Σa_n is convergent.

It follows from this theorem that if $a_n \geqslant 0$, $b_n \geqslant 0$, if $\lim a_n/b_n = l \geqslant 0$ and if Σb_n is convergent then Σa_n is also convergent.

THEOREM 19. *If $a_n \geqslant 0$, $b_n \geqslant 0$, if there is a positive number k independent of n and an integer N such that $a_n > kb_n$ whenever $n > N$, and if Σb_n is properly divergent then Σa_n is also properly divergent.*

If $n > N$ we have

$$A_n - A_N = \sum_{\nu=N+1}^{n} a_\nu > k \sum_{\nu=N+1}^{n} b_\nu = k(B_n - B_N),$$

whence

$$A_n > kB_n + A_N - kB_N.$$

Let $n \to \infty$. Then $A_n \to \infty$ since $B_n \to \infty$ and $k > 0$.

The reader should satisfy himself that it is possible to construct a proof of Theorem 18 along the lines of the proof of Theorem 19 and a proof of Theorem 19 along the lines of the proof of Theorem 18. It should be noted that in Theorem 19 it is essential that k be *greater* than zero.

As in the case of Theorem 18 it follows that if $a_n \geqslant 0$, $b_n \geqslant 0$, if $\lim a_n/b_n = l > 0$ and if Σb_n is properly divergent then Σa_n is properly divergent.

In the case when the hypotheses of both theorems are satisfied we have the following theorem.

THEOREM 20. *If $a_n \geqslant 0$, $b_n \geqslant 0$, if positive numbers k and K, independent of n, and a positive integer N can be found such that, whenever $n > N$,*

$$k < a_n / b_n < K,$$

then Σa_n is convergent or properly divergent according as Σb_n is convergent or properly divergent.

In this case we say that Σa_n and Σb_n converge or properly diverge together or that Σa_n behaves like Σb_n.

We note that, in particular, the conclusion of the theorem will be true if $\lim a_n / b_n = l > 0$.

Example 1.

The series $\sum_{n=2}^{\infty} \frac{1}{n(\log n)^{\lambda}}, \quad \sum_{n=3}^{\infty} \frac{1}{n \log n (\log \log n)^{\lambda}}, \ldots$

are properly divergent for $\lambda < 0$.

This follows at once from Theorem 19 by comparing these series respectively with the following properly divergent series :—

$$\sum_{n=2}^{\infty} \frac{1}{n}, \quad \sum_{n=3}^{\infty} \frac{1}{n \log n}, \ldots .$$

Example 2. Examine the convergence of the series $\sum_{n=1}^{\infty} \log (1+n^{-\lambda})$.

The series is properly divergent for $\lambda \leqslant 0$ since its n-th term does not tend to zero. When $\lambda > 0$,

$$\lim_{n \to \infty} \frac{\log (1+n^{-\lambda})}{n^{-\lambda}} = 1.$$

Thus the series behaves like $\Sigma n^{-\lambda}$; that is, it is convergent for $\lambda > 1$ and properly divergent for $0 < \lambda \leqslant 1$.

Example 3. Examine for convergence the series Σa_n where

$$a_n = \frac{2^{1+(-1)^n}}{n^2+p^2}.$$

We have $a_n < 4/n^2$ and, for $n \geqslant |p|$, $a_n > 1/(2n^2)$. The series therefore behaves like $\Sigma 1/n^2$; that is, it is convergent.

26. The Ratio or d'Alembert's Test.

THEOREM 21. *If* $a_n > 0$ *and if* $\lim \dfrac{a_{n+1}}{a_n} = \rho$ *then* Σa_n *is convergent if* $\rho < 1$ *and properly divergent if* $\rho > 1$.

Suppose first that $\rho < 1$. Then given $\epsilon (< 1 - \rho)$ we can find $N = N(\epsilon)$ such that, whenever $n \geqslant N$,

$$a_{n+1} < (\rho + \epsilon) a_n.$$

In particular,

$$\begin{aligned} &a_{N+1} < (\rho+\epsilon) a_N, \\ &a_{N+2} < (\rho+\epsilon) a_{N+1} < (\rho+\epsilon)^2 a_N, \\ &\qquad \cdots\cdots \\ &a_{N+m} < (\rho+\epsilon) a_{N+m-1} < (\rho+\epsilon)^m a_N. \end{aligned}$$

Since $0 < \rho + \epsilon < 1$ the series $\sum_{m=1}^{\infty} (\rho+\epsilon)^m a_N$ is convergent. It follows from Theorem 18 that $\sum_{\nu=N+1}^{\infty} a_\nu$ is convergent and therefore that $\sum_{\nu=1}^{\infty} a_\nu$ is convergent.

Suppose next that $\rho > 1$. Given $\epsilon (< \rho - 1)$ we can find $N = N(\epsilon)$ such that, whenever $n \geqslant N$,

$$a_{n+1} > (\rho - \epsilon) a_n.$$

It follows, as in the first part of the proof, that, for $m \geqslant 1$,

$$a_{N+m} > (\rho - \epsilon)^m a_N,$$

and the result follows from Theorem 19.

We note in passing that when $\rho = 1$ the test yields no definite conclusion. For example, in the case of the series $\Sigma n^{-\lambda}$ it is easy to see that $\rho = 1$ no matter what value λ may have. The series is only convergent, however, when $\lambda > 1$.

It may happen in the case of some series that $\lim a_{n+1}/a_n$ does not exist. It is clear from the above proofs, however, that such series will be convergent if $\overline{\lim}\, a_{n+1}/a_n < 1$ and properly divergent if $\underline{\lim}\, a_{n+1}/a_n > 1$.

Example. Examine for convergence the series $\sum_{n=1}^{\infty} n^{\lambda} x^n$, $(x > 0)$. For this series *

$$\frac{a_{n+1}}{a_n} = \left(1 + \frac{1}{n}\right)^{\lambda} x \to x.$$

Thus the series is convergent for $0 < x < 1$ and any value of λ. It is properly divergent for $x > 1$ and any value of λ. When $x = 1$ the series becomes Σn^{λ} and the behaviour of this series has already been discussed.

27. Cauchy's Test.

THEOREM 22. *If* $a_n \geqslant 0$ *and if*

$$\overline{\lim}\, \sqrt[n]{a_n} = \rho,$$

then Σa_n *is convergent if* $\rho < 1$ *and properly divergent if* $\rho > 1$.

Suppose first that $\rho < 1$. Given $\epsilon (< 1 - \rho)$ we can find $N = N(\epsilon)$ such that $\sqrt[n]{a_n} < \rho + \epsilon$, that is, $a_n < (\rho + \epsilon)^n$ whenever $n \geqslant N$. It follows from Theorem 18 that Σa_n is convergent since $0 < \rho + \epsilon < 1$.

Suppose now that $\rho > 1$. Given $\epsilon (< \rho - 1)$ we can find an infinity of values of n, say $n_1, n_2, \ldots$ such that, for these values of n, $a_n > (\rho - \epsilon)^n$. Since $\rho - \epsilon > 1$ it follows that a_n cannot tend to zero so that the series Σa_n is properly divergent.

A more common but less general form of the theorem is obtained by replacing $\overline{\lim}\, \sqrt[n]{a_n}$ by $\lim \sqrt[n]{a_n}$. As in the case of the Ratio test no conclusion can be drawn when $\rho = 1$ for, considering again the series $\Sigma n^{-\lambda}$, we have

$$\log \sqrt[n]{a_n} = -\frac{\lambda}{n} \log n \to 0,$$

so that, for all values of λ, $\sqrt[n]{a_n} \to 1$.

* The series is obviously convergent when $x = 0$.

28. Connection between the Ratio Test and Cauchy's Test. We shall now show that Cauchy's test is more general than the Ratio test.

THEOREM 23. *If* $a_n>0$ *and if* a_{n+1}/a_n *tends to a limit then* $\sqrt[n]{a_n}$ *tends to the same limit.*

Suppose that $a_{n+1}/a_n \to \rho$ where ρ is finite and not zero. Then $\log a_{n+1} - \log a_n \to \log \rho$. That is, given ϵ, we can find $N = N(\epsilon)$ such that, whenever $n \geqslant N$,

$$\log \rho - \epsilon < \log a_{n+1} - \log a_n < \log \rho + \epsilon.$$

For n write in turn N, $N+1$, ... $N+m-1$, and add. Then, for $m \geqslant 1$,

$$m(\log \rho - \epsilon) < \log a_{N+m} - \log a_N < m(\log \rho + \epsilon),$$

that is,

$$\log \rho - \epsilon < (1/m) \log a_{N+m} - (1/m) \log a_N < \log \rho + \epsilon.$$

Let $m \to \infty$. Then

$$\log \rho - \epsilon \leqslant \varliminf_{m\to\infty} (1/m) \log a_{N+m} \leqslant \varlimsup_{m\to\infty} (1/m) \log a_{N+m} \leqslant \log \rho + \epsilon.$$

Since ϵ is arbitrary it follows that $\lim\limits_{m\to\infty} \dfrac{1}{m} \log a_{N+m}$ exists and is equal to $\log \rho$. Hence writing $\nu = N+m$ we see that $\lim\limits_{\nu\to\infty} \dfrac{1}{\nu} \log a_\nu = \log \rho$; that is $\sqrt[\nu]{a_\nu} \to \rho$.

The proof only requires trivial alterations in order to be applicable also to the cases when ρ is zero or infinite.

Exactly the same method of proof suffices to show that if

$$\varliminf a_{n+1}/a_n = \omega, \quad \varlimsup a_{n+1}/a_n = \Omega,$$

then

$$\omega \leqslant \varliminf \sqrt[n]{a_n} \leqslant \varlimsup \sqrt[n]{a_n} \leqslant \Omega.$$

We have therefore shown that, whenever a series can be proved to be convergent or properly divergent by the

Ratio test it can also be proved convergent or properly divergent by Cauchy's test. We shall now give an example to show that there are series for which a direct application of the Ratio test gives no result but whose behaviour may be determined by Cauchy's test.

Consider the series Σa_n where

$$a_n = 2^{-n-(-1)^n}.$$

Clearly

$$\sqrt[n]{a_n} = 2^{-1-(-1)^n/n} \to \tfrac{1}{2},$$

so that, by Cauchy's test, the series is convergent. Also

$$\frac{a_{n+1}}{a_n} = 2^{-1+(-1)^n-(-1)^{n+1}}$$

which is 2 if n is even and $\frac{1}{8}$ if n is odd. Thus $\overline{\lim}\, a_{n+1}/a_n = 2$ and $\underline{\lim}\, a_{n+1}/a_n = \frac{1}{8}$, so that the Ratio test yields no definite result.

29. A General Test for Convergence. We have seen that, in cases when the ratio a_{n+1}/a_n tends to unity, no conclusion can be drawn as regards the behaviour of the series Σa_n. The tests which we shall discuss in this and the subsequent articles are more delicate than the Ratio test and enable us to arrive at a conclusion in such cases. These tests are particular cases of a general test due to Kummer, which we now proceed to obtain.

THEOREM 24. *Suppose that $a_n>0$, $b_n>0$ and that Σb_n is properly divergent. Let*

$$\lim \left(\frac{1}{b_n}\frac{a_n}{a_{n+1}} - \frac{1}{b_{n+1}}\right) = \kappa.$$

Then Σa_n converges or properly diverges according as $\kappa>0$ or $\kappa<0$.

Suppose first that $\kappa>0$. Then we can find N such that, whenever $n \geqslant N$,

$$\frac{1}{b_n}\frac{a_n}{a_{n+1}} - \frac{1}{b_{n+1}} > \tfrac{1}{2}\kappa;$$

that is,

$$a_{n+1} < \frac{2}{\kappa}\left(\frac{a_n}{b_n} - \frac{a_{n+1}}{b_{n+1}}\right).$$

Thus

$$\sum_{\nu=N+1}^{n+1} a_\nu < \frac{2}{\kappa}\left(\frac{a_N}{b_N} - \frac{a_{n+1}}{b_{n+1}}\right)$$

$$< \frac{2}{\kappa}\frac{a_N}{b_N}.$$

Hence, for the series Σa_n, A_n is bounded, so that the series is convergent.

Next suppose that $\kappa < 0$. Then we can find N such that, whenever $n \geqslant N$,

$$\frac{1}{b_n}\frac{a_n}{a_{n+1}} - \frac{1}{b_{n+1}} < 0;$$

that is,

$$a_{n+1} > \frac{a_n}{b_n} b_{n+1}.$$

In particular,

$$a_{N+1} > \frac{a_N}{b_N} b_{N+1},$$

$$a_{N+2} > \frac{a_{N+1}}{b_{N+1}} b_{N+2} > \frac{a_N}{b_N} b_{N+2},$$

$$\cdots\cdot$$

$$a_{N+m} > \frac{a_N}{b_N} b_{N+m}.$$

The result then follows from Theorem 19 since Σb_n is properly divergent.

It is clear that in cases when $\lim\left(\frac{1}{b_n}\frac{a_n}{a_{n+1}} - \frac{1}{b_{n+1}}\right)$ does not exist the series Σa_n will be convergent if $\underline{\lim}\left(\frac{1}{b_n}\frac{a_n}{a_{n+1}} - \frac{1}{b_{n+1}}\right) > 0$ and will be properly divergent if $\overline{\lim}\left(\frac{1}{b_n}\frac{a_n}{a_{n+1}} - \frac{1}{b_{n+1}}\right) < 0$.

It should also be noted that the Ratio test corresponds to the particular case $b_n = 1$, $(n = 1, 2, \ldots)$ of Theorem 24.

30. Raabe's Test.

THEOREM 25. *Suppose that $a_n > 0$ and that, as $n \to \infty$,*

$$\frac{a_n}{a_{n+1}} = 1 + \frac{\sigma}{n} + o\left(\frac{1}{n}\right).$$

Then Σa_n is convergent or properly divergent according as $\sigma > 1$ or $\sigma < 1$.

From the hypothesis

$$\lim \left\{ n \frac{a_n}{a_{n+1}} - (n+1) \right\} = \sigma - 1,$$

so that the result follows from Theorem 24 on writing $\kappa = \sigma - 1$, $b_n = n^{-1}$, $(n = 1, 2, \ldots)$.

Example.—Examine for convergence the series

$$\sum_{n=1}^{\infty} \frac{1.3.5 \ldots (2n-1)}{2.4.6 \ldots 2n} \cdot \frac{1}{n}.$$

For this series

$$\frac{a_n}{a_{n+1}} = \frac{(2n+2)(n+1)}{(2n+1)n} = \frac{2n^2+4n+2}{2n^2+n}$$

$$= 1 + \frac{3}{2n} + O\left(\frac{1}{n^2}\right),$$

and it follows at once from Raabe's test that the series is convergent.

It should be noted that in this case the Ratio test gives no information.

31. Gauss's Test. We have seen that Raabe's test gives no information when $\sigma = 1$. Gauss's test is a slight modification of Raabe's test which usually enables us to settle the case $\sigma = 1$ without having recourse to a separate argument. We require first a further deduction from Theorem 24.

Theorem 26. *If $a_n > 0$ and if*

$$\frac{a_n}{a_{n+1}} = 1 + \frac{1}{n} + \frac{\tau_n}{n \log n}$$

then Σa_n is convergent or properly divergent according as $\lim \tau_n$ is greater than or less than unity.

Put $b_n = \dfrac{1}{n \log n}$, $(n = 2, 3, \ldots)$ in Theorem 24. Then

$$\frac{1}{b_n}\frac{a_n}{a_{n+1}} - \frac{1}{b_{n+1}} = n \log n \frac{a_n}{a_{n+1}} - (n+1) \log (n+1)$$
$$= \omega_n - 1,$$

say. The series Σa_n will be convergent or properly divergent according as $\lim \omega_n$ is greater than or less than unity The last identity may also be written

$$\frac{a_n}{a_{n+1}} = \frac{n+1}{n} \frac{\log (n+1)}{\log n} + \frac{\omega_n - 1}{n \log n}$$

$$= \left(1 + \frac{1}{n}\right)\left\{1 + \frac{\log\left(1 + \frac{1}{n}\right)}{\log n}\right\} + \frac{\omega_n - 1}{n \log n}$$

$$= \left(1 + \frac{1}{n}\right)\left\{1 + \frac{1}{n \log n} + O\left(\frac{1}{n^2 \log n}\right)\right\} + \frac{\omega_n - 1}{n \log n}$$

$$= 1 + \frac{1}{n} + \frac{\omega_n}{n \log n} + O\left(\frac{1}{n^2 \log n}\right)$$

$$= 1 + \frac{1}{n} + \frac{\omega_n + o(1)}{n \log n}$$

and from this the result follows.

Theorem 27. *If $a_n > 0$ and if*

$$\frac{a_n}{a_{n+1}} = 1 + \frac{\sigma}{n} + O\left(\frac{1}{n^{\delta+1}}\right), \; \delta > 0,$$

then Σa_n converges if $\sigma > 1$ and is properly divergent if $\sigma \leqslant 1$.

From the hypothesis

$$\frac{a_n}{a_{n+1}} = 1+\frac{\sigma}{n}+o\left(\frac{1}{n}\right),$$

so that, by Raabe's test, the series is convergent for $\sigma>1$ and properly divergent for $\sigma<1$. When $\sigma = 1$

$$\frac{a_n}{a_{n+1}} = 1+\frac{1}{n}+\frac{O(n^{-\delta}\log n)}{n\log n}$$
$$= 1+\frac{1}{n}+\frac{o(1)}{n\log n}$$

so that Σa_n is properly divergent by Theorem 26.

Example. Examine for convergence the series

$$1+\frac{\alpha\,.\,\beta}{1\,.\,\gamma}+\frac{\alpha(\alpha+1)\beta(\beta+1)}{1\,.\,2\,.\,\gamma(\gamma+1)}+\ldots,$$

where α, β, γ are neither zero nor negative integers.

Clearly the terms of this series are ultimately of the same sign. For convenience we denote the first term of the series by a_0 instead of by a_1. We then have

$$\frac{a_n}{a_{n+1}} = \frac{(n+1)(\gamma+n)}{(\alpha+n)(\beta+n)} = \frac{n^2+(\gamma+1)n+\gamma}{n^2+(\alpha+\beta)n+\alpha\beta}$$
$$= 1+\frac{\gamma+1-\alpha-\beta}{n}+O\left(\frac{1}{n^2}\right).$$

Thus the series is convergent if $\gamma>\alpha+\beta$ and is properly divergent if $\gamma\leqslant\alpha+\beta$.

When α is replaced by $-\alpha$ and $\beta = \gamma$, the above series becomes

$$1+\frac{\alpha}{1}(-1)+\frac{\alpha(\alpha-1)}{1\,.\,2}(-1)^2+\ldots,$$

that is, the binomial series when $x = -1$. Our argument shows that it is convergent when $\alpha>0$ and properly divergent when $\alpha<0$. When $\alpha = 0$ the series becomes $1+0+0+\ldots$ which is convergent.

32. Euler's Constant. We conclude this chapter with two very important relations. The first is a direct deduction from Theorem 16. If $f(x) = x^{-1}$, $(x>0)$, Theorem 16 shows that

$$1+\frac{1}{2}+\frac{1}{3}+\ldots+\frac{1}{n}-\log n$$

tends to a limit γ which is such that $0 \leqslant \gamma \leqslant 1$. This number γ is called Euler's constant and its value is 0·57721566... The following rather less precise result is an obvious corollary:

$$\sum_{\nu=1}^{n} \nu^{-1} \sim \log n.$$

Example.—

Evaluate $\sum\limits_{n=1}^{\infty} \dfrac{1}{n(2n+1)}$.

Since $\dfrac{1}{n(2n+1)} = \dfrac{1}{n} - \dfrac{2}{2n+1}$ we have for this series

$$A_n = \left(1+\frac{1}{2}+\ldots+\frac{1}{n}\right) - 2\left(\frac{1}{3}+\frac{1}{5}+\ldots+\frac{1}{2n+1}\right)$$

$$= 2-\frac{2}{2n+1}+\left(1+\frac{1}{2}+\ldots+\frac{1}{n}\right)-2\left(1+\frac{1}{3}+\ldots+\frac{1}{2n-1}\right)$$

$$= 2-\frac{2}{2n+1}+2\left(1+\frac{1}{2}+\ldots+\frac{1}{n}\right)-2\left(1+\frac{1}{2}+\frac{1}{3}+\ldots+\frac{1}{2n}\right)$$

$$= 2+O\left(\frac{1}{n}\right)+2\{\log n+\gamma+o(1)\}-2\{\log 2n+\gamma+o(1)\}$$

$$= 2-2\log 2+o(1).$$

Hence

$$\sum_{n=1}^{\infty} \frac{1}{n(2n+1)} = 2-2\log 2.$$

33. Stirling's Approximation for $n!$ We shall now prove that, as $n \to \infty$,

$$n! \sim \sqrt{(2\pi)}\, n^{n+\frac{1}{2}} e^{-n}.$$

In the proof of this result we require the value of the integral

$$T = \int_0^{\frac{1}{2}} \log\left(\frac{\sin \pi t}{\pi t}\right) dt,$$

and we shall obtain this first.

We have, on putting $\pi t = \theta$,

$$\begin{aligned}\pi T &= \int_0^{\frac{1}{2}\pi} \log\left(\frac{\sin\theta}{\theta}\right) d\theta \\ &= \int_0^{\frac{1}{2}\pi} \log \sin\theta\, d\theta - \int_0^{\frac{1}{2}\pi} \log\theta\, d\theta \\ &= T_1 - [\theta\log\theta - \theta]_{\theta\to 0}^{\theta=\frac{1}{2}\pi} \\ &= T_1 + \tfrac{1}{2}\pi - \tfrac{1}{2}\pi \log \tfrac{1}{2}\pi.\end{aligned}$$

Now T_1 is convergent (see Art. **10**) and

$$T_1 = \int_0^{\frac{1}{2}\pi} \log \sin\theta\, d\theta = \int_0^{\frac{1}{2}\pi} \log \cos\theta\, d\theta,$$

whence

$$\begin{aligned}2T_1 &= \int_0^{\frac{1}{2}\pi} \log\left(\tfrac{1}{2}\sin 2\theta\right) d\theta \\ &= -\tfrac{1}{2}\pi \log 2 + \int_0^{\frac{1}{2}\pi} \log \sin 2\theta\, d\theta \\ &= -\tfrac{1}{2}\pi \log 2 + \tfrac{1}{2}\int_0^{\pi} \log \sin\phi\, d\phi, \quad (\phi = 2\theta), \\ &= -\tfrac{1}{2}\pi \log 2 + T_1.\end{aligned}$$

Thus

$$T_1 = -\tfrac{1}{2}\pi \log 2,$$

and

$$\pi T = \tfrac{1}{2}\pi - \tfrac{1}{2}\pi \log 2 - \tfrac{1}{2}\pi \log \pi + \tfrac{1}{2}\pi \log 2,$$

so that

$$T = \tfrac{1}{2} - \tfrac{1}{2}\log \pi.$$

We now proceed to prove the main result.

We have

$$\log \nu = \int_{-\frac{1}{2}}^{\frac{1}{2}} \log \nu \, dt = \int_{-\frac{1}{2}}^{\frac{1}{2}} \log (\nu+t) dt - \int_{-\frac{1}{2}}^{\frac{1}{2}} \log \left(1+\frac{t}{\nu}\right) dt$$

$$= \int_{\nu-\frac{1}{2}}^{\nu+\frac{1}{2}} \log t \, dt - \int_{0}^{\frac{1}{2}} \log \left(1+\frac{t}{\nu}\right) dt - \int_{-\frac{1}{2}}^{0} \log \left(1+\frac{t}{\nu}\right) dt$$

$$= \int_{\nu-\frac{1}{2}}^{\nu+\frac{1}{2}} \log t \, dt - \int_{0}^{\frac{1}{2}} \log \left(1-\frac{t^2}{\nu^2}\right) dt,$$

whence

$$\log (n!) = \sum_{\nu=1}^{n} \log \nu = \int_{\frac{1}{2}}^{n+\frac{1}{2}} \log t \, dt - \int_{0}^{\frac{1}{2}} \left\{ \sum_{\nu=1}^{n} \log \left(1-\frac{t^2}{\nu^2}\right) \right\} dt.$$

The series $\sum_{\nu=1}^{\infty} -\log \left(1-\frac{t^2}{\nu^2}\right)$ behaves, for $0<t\leqslant\frac{1}{2}$, like the series $\sum_{\nu=1}^{\infty} +1/\nu^2$; that is, it is convergent. Its sum (see Art. 53) is $-\log \left(\frac{\sin \pi t}{\pi t}\right)$. Hence

$$\log (n!) = \int_{\frac{1}{2}}^{n+\frac{1}{2}} \log t \, dt - \int_{0}^{\frac{1}{2}} \log \left(\frac{\sin \pi t}{\pi t}\right) dt$$

$$+ \int_{0}^{\frac{1}{2}} \left\{ \sum_{\nu=n+1}^{\infty} \log \left(1-\frac{t^2}{\nu^2}\right) \right\} dt$$

$$= T_1(n)+T_2(n)+T_3(n),$$

say. Now

$$T_1(n) = \Big[t \log t - t\Big]_{\frac{1}{2}}^{n+\frac{1}{2}} = (n+\tfrac{1}{2}) \log (n+\tfrac{1}{2}) - n + \tfrac{1}{2} \log 2,$$

$$T_2(n) = \tfrac{1}{2} \log \pi - \tfrac{1}{2},$$

and

$$|T_3(n)| = \int_{0}^{\frac{1}{2}} \left\{ \sum_{\nu=n+1}^{\infty} \log \left(1 \Big/ \left(1-\frac{t^2}{\nu^2}\right)\right) \right\} dt$$

$$\leqslant \tfrac{1}{2} \sum_{\nu=n+1}^{\infty} \log \left(1 \Big/ \left(1-\frac{1}{4\nu^2}\right)\right) = O\left(\sum_{\nu=n+1}^{\infty} \frac{1}{\nu^2}\right)$$

$$= O\left(\int_{n+1}^{\infty} \frac{1}{x^2} dx\right) = O\left(\frac{1}{n}\right).$$

Collecting these results we obtain

$$\log (n!) = \left(n+\frac{1}{2}\right) \log n + \left(n+\frac{1}{2}\right) \log \left(1+\frac{1}{2n}\right)$$
$$-n-\tfrac{1}{2}+\tfrac{1}{2}\log 2\pi+O\left(\frac{1}{n}\right)$$
$$= \left(n+\frac{1}{2}\right) \log n - n + \tfrac{1}{2} \log 2\pi + O\left(\frac{1}{n}\right),$$

from which the result follows.

Although the result stated at the beginning of the article is sufficient for most applications it should be noted that we have really obtained something more precise. We have in fact proved that, as $n \to \infty$,

$$n! = \sqrt{(2\pi)} n^{n+\frac{1}{2}} e^{-n} \left\{1+O\left(\frac{1}{n}\right)\right\}.$$

Examples

1. If Σa_n converges to the sum a, prove that $B_n \sim a \log n$ where

$$b_n = \frac{1}{n} (a_1+a_2+\ldots+a_n).$$

2. If $a_n \sim an^\rho$, $(\rho > -1)$, prove that $A_n \sim \dfrac{an^{\rho+1}}{\rho+1}$.

3. Prove that, if $\rho < 1$,

$$\frac{1}{(n+1)^\rho}+\frac{1}{(n+2)^\rho}+\ldots+\frac{1}{(2n)^\rho} \sim \frac{2^{1-\rho}-1}{1-\rho} n^{1-\rho}.$$

4. Prove that, as $n \to \infty$,

$$\frac{1}{2 \log 2}+\frac{1}{3 \log 3}+\ldots+\frac{1}{n \log n} - \log (\log n)$$

tends to a definite limit. Deduce that, if p is a positive integer,

$$\lim_{n\to\infty} \sum_{\nu=n}^{n^p} \frac{1}{\nu \log \nu} = \log p.$$

5. Examine for convergence the series

(i) $\Sigma \dfrac{1}{n^2+a^2}$, (ii) $\Sigma \dfrac{a-bn}{a+bn^2}$, (iii) $\Sigma \dfrac{1}{n^p+a}$, (iv) $\Sigma \dfrac{1}{2^n+x}$,

(v) Σe^{-n^2x}, (vi) $\Sigma \dfrac{1}{\sqrt{n}+\sqrt{(n+1)}}$, (vii) $\Sigma \dfrac{n}{\sqrt{(2n^3+1)}}$,

(viii) $\Sigma \sqrt{\left(\dfrac{n}{n^4+1}\right)}$, (ix) $\Sigma \dfrac{\sqrt{(x+n)}-1}{\sqrt{(x^2+n^2)}+1}$,

(x) $\Sigma \dfrac{1}{n} \{\sqrt{(n^2+n+1)}-\sqrt{(n^2-n+1)}\}$,

(xi) $\Sigma \dfrac{1}{n} \{\sqrt{(n+1)}-\sqrt{(n-1)}\}$, (xii) $\sum\limits_{n=2}^{\infty} \dfrac{1}{(\log n)^{\alpha}}$,

(xiii) $\Sigma \dfrac{n^{\alpha}}{n!}$, (xiv) $\Sigma \dfrac{n^n}{n!}$, (xv) $\Sigma \dfrac{(n!)^2}{(2n)!}\, x^{2n}$,

(xvi) $\Sigma \dfrac{n!}{x(x+1)\ldots(x+n-1)}$, (xvii) $\Sigma \dfrac{\alpha(\alpha+1)\ldots(\alpha+n-1)}{n^n}$,

(xviii) $\Sigma \sqrt{\left\{\dfrac{\alpha(\alpha+1)\ldots(\alpha+n-1)}{\beta(\beta+1)\ldots(\beta+n-1)}\right\}}$, (xix) $\Sigma \left\{\dfrac{(2n)!}{2^{2n}(n!)^2}\right\}^{\alpha}$,

(xx) $\Sigma \dfrac{1}{n \sqrt[n]{n}}$, (xxi) $\Sigma \left\{n \log \dfrac{3n+2}{3n-2} - 1\right\}$,

(xxii) $\Sigma \left(\sin \dfrac{x}{n}\right)^{\alpha}$, (xxiii) $\Sigma \dfrac{1}{n^{\alpha}}\left(1+\dfrac{1}{2^{\alpha}}+\ldots+\dfrac{1}{n^{\alpha}}\right)$,

(xxiv) $\Sigma \left(\dfrac{n}{3n+1}\right)^{n^3}$, (xxv) $\sum\limits_{n=2}^{\infty} \dfrac{1}{(\log n)^{2n}}$,

(xxvi) $\sum\limits_{n=2}^{\infty} \dfrac{1}{(\log n)^{\log n}}$, (xxvii) $\sum\limits_{n=3}^{\infty} \dfrac{1}{(\log \log n)^{\log n}}$,

(xxviii) $\sum\limits_{n=3}^{\infty} \dfrac{1}{(\log n)^{\log \log n}}$, (xxix) $\Sigma \dfrac{1.3.5\ldots 2n-1}{2.4.6\ldots.2n} \cdot \dfrac{1}{n^{\alpha}}$,

(xxx) $\sum\limits_{n=2}^{\infty} \left\{1+\dfrac{1}{n(\log n)^{\lambda}}\right\}^{-n^{\alpha}}$,

(xxxi)

$$\Sigma \frac{\alpha(\alpha+1)\ldots(\alpha+n-1)\beta(\beta+1)\ldots(\beta+n-1)\gamma(\gamma+1)\ldots(\gamma+n-1)}{1.2\ldots n\, \delta(\delta+1)\ldots(\delta+n-1)\zeta(\zeta+1)\ldots(\zeta+n-1)}.$$

6. If $a_n = \dfrac{1}{3n-2}+\dfrac{1}{3n-1}-\dfrac{1}{3n}$

prove that

$$A_n = \tfrac{1}{3}\log n+\log 3+\tfrac{1}{3}\gamma+o(1).$$

7. Show that

$$\sum_{n=1}^{\infty}\frac{1}{n(4n^2-1)} = 2\log 2-1 \text{ and that } \sum_{r=0}^{n}\frac{1}{n+r}\to\log 2.$$

8. If $A_n = 1+\dfrac{1}{2}+\ldots+\dfrac{1}{n}$, prove that

(i) $1+\tfrac{1}{2}n\leqslant A_{2^n}\leqslant n+(\tfrac{1}{2})^n$, (ii) $\sqrt[n]{A_n}\to 1$.

9. Prove that

$$\sum_{r=1}^{n}\frac{n-r+1}{r}\sim\log(n!).$$

[Use Example 13, p. 35.]

10. Prove that

(i) $\lim \dfrac{n+1}{(n!)^{1/n}} = e,$

(ii) $\lim \dfrac{\{(n+1)(n+2)\ldots(n+n)\}^{1/n}}{n} = 4/e.$

[Use Theorem 23.]

11. Prove that

$$\lim \frac{1}{n}\,[(n^2+1^2)(n^2+2^2)^2\ldots(n^2+n^2)^n]^{1/n^2} = 2/\sqrt{e}.$$

ANSWERS. 5. (i) Convergent; (ii) properly divergent; (iii) convergent if $p>1$, properly divergent if $p\leqslant 1$; (iv) convergent; (v) convergent if $x>0$, properly divergent if $x\leqslant 0$; (vi) properly divergent; (vii) properly divergent; (viii) convergent; (ix) properly divergent; (x) properly divergent; (xi) convergent; (xii) properly divergent; (xiii) convergent; (xiv) properly divergent; (xv) convergent if $|x|<2$, properly divergent if $|x|\geqslant 2$; (xvi) convergent if $x>2$, properly divergent if $x\leqslant 2$, zero and negative integral values being excluded;

(xvii) convergent ; (xviii) convergent if $\beta-\alpha>2$, properly divergent if $\beta-\alpha\leqslant 2$, zero and negative integral values of α and β being excluded ; (xix) convergent if $\alpha>2$, properly divergent if $\alpha\leqslant 2$; (xx) properly divergent ; (xxi) properly divergent ; (xxii) convergent if $\alpha>1$, properly divergent when $\alpha\leqslant 1$, provided that $x\neq 0$; (xxiii) convergent if $\alpha>1$, properly divergent if $\alpha\leqslant 1$; (xxiv) convergent ; (xxv) convergent ; (xxvi) convergent ; (xxvii) convergent ; (xxviii) properly divergent ; (xxix) convergent if $\alpha>\frac{1}{2}$, properly divergent if $\alpha\leqslant\frac{1}{2}$; (xxx) convergent if $\alpha>1$, properly divergent if $\alpha<1$, while, if $\alpha = 1$, convergent for $\lambda<-1$, properly divergent for $\lambda\geqslant -1$; (xxxi) convergent if $\delta+\zeta-\alpha-\beta-\gamma>0$, otherwise properly divergent, the constants being such that none of the factors vanishes.

CHAPTER V

GENERAL SERIES

34. Real Series. We turn now from the special case of series whose terms are all of the same sign to series whose terms may be real and of either sign and to series whose terms may be complex. We consider first real series.

35. Absolute Convergence. Before defining what we mean by absolute convergence we prove the following theorem.

THEOREM 28. *If the series $\Sigma|a_n|$ is convergent, then so is the series Σa_n.*

Let
$$u_n = a_n\,,\ (a_n \geqslant 0) \qquad v_n = -a_n\,,\ (a_n \leqslant 0),$$
$$\quad = 0\,,\ (a_n \leqslant 0)\ ; \qquad\ = 0\,,\ (a_n \geqslant 0).$$

Then clearly $u_n \geqslant 0$, $v_n \geqslant 0$ and

$$|a_n| = u_n + v_n\,,\ a_n = u_n - v_n.$$

From the first of these relations it follows that

$$u_n \leqslant |a_n|\,,\ v_n \leqslant |a_n|.$$

Since $\Sigma|a_n|$ is convergent, both Σu_n and Σv_n are convergent by Theorem 18. Hence, by Theorem 10, $\Sigma(u_n - v_n)$ is convergent; that is, Σa_n is convergent.

The proper divergence of $\Sigma|a_n|$ does not imply the divergence of Σa_n. For example, if $a_n = (-1)^{n-1}n^{-1}$ we have seen that $\sum_{n=1}^{\infty}|a_n|$ is properly divergent, whereas $\sum_{n=1}^{\infty} a_n$ converges to the sum log 2.

If Σa_n is a series such that $\Sigma|a_n|$ is convergent, then we say that Σa_n is **absolutely convergent.** Theorem 28, therefore, merely states that every absolutely convergent series is necessarily convergent. From the definition it is perhaps reasonable to expect that absolutely convergent series should possess many of the properties of series whose terms are non-negative. We have an instance of this in the following theorem.

THEOREM 29. *If Σa_n is an absolutely convergent series and if Σb_n is a series whose terms are those of Σa_n in a different order then Σb_n is absolutely convergent and the sums of the two series are the same.*

Define u_n and v_n as in Theorem 28 and let u_n', v_n' be defined in a similar way for the series Σb_n. Since Σa_n is absolutely convergent the series Σu_n and Σv_n are convergent series of non-negative terms. It is clear that $\Sigma u_n'$ and $\Sigma v_n'$ are formed from Σu_n and Σv_n respectively simply by an alteration in the order of the terms. Hence, by Theorem 14, $\Sigma u_n'$ and $\Sigma v_n'$ converge respectively to the sums of the series Σu_n and Σv_n. Thus $\Sigma b_n = \Sigma(u_n'-v_n')$ is convergent to the sum of the series $\Sigma(u_n-v_n) = \Sigma a_n$.

The *absolute* convergence of Σb_n follows at once from Theorem 14 since $\Sigma|a_n|$ is convergent.

36. Tests for Absolute Convergence. The question of examining whether or not a series Σa_n is absolutely convergent resolves itself into testing for convergence the series of non-negative terms $\Sigma|a_n|$. This may be done by making use of the tests given in Chapter IV. It should be observed, however, that if

$$\lim \frac{|a_{n+1}|}{|a_n|} > 1 \text{ or } \overline{\lim} \sqrt[n]{|a_n|} > 1,$$

the series Σa_n is not merely not absolutely convergent but is in fact divergent. This follows from the fact that each of the above conditions implies that $a_n \not\to 0$, so that the necessary condition, $a_n \to 0$, for the convergence of the series Σa_n is not satisfied.

The series $1-\frac{1}{2}+\frac{1}{3}-\ldots$ is convergent but not absolutely convergent since the series $1+\frac{1}{2}+\frac{1}{3}+\ldots$ is not convergent. A series which is convergent but not absolutely convergent is said to be **conditionally convergent.**

Example. Examine the convergence of the series

$$\sum_{n=1}^{\infty}\{\log(n+1)\}^{a}x^{n}.$$

For this series we have

$$\frac{|a_{n+1}|}{|a_n|}=\left\{\frac{\log(n+2)}{\log(n+1)}\right\}^{a}.\ |x|\to|x|.$$

Thus the series is absolutely convergent for $-1<x<1$ and divergent for $x>1$ and for $x<-1$. When $x=1$ the series is properly divergent. When $x=-1$ and $a\geqslant 0$ the series is divergent since its nth term does not tend to zero. When $x=-1$ and $a<0$ the series may be shown to be conditionally convergent (see Theorem 32 below).

37. Conditional Convergence. We now consider real series which are convergent but not absolutely convergent. We shall obtain tests for the convergence of such series which are of wide application. First we prove a subsidiary lemma.

LEMMA. *If b_n is a positive, monotonic decreasing function and if A_n is bounded, then the series $\Sigma A_n(b_n-b_{n+1})$ is absolutely convergent.*

Suppose that $|A_n|<K$. Then

$$\begin{aligned}\sum_{n=1}^{N}|A_n(b_n-b_{n+1})| &= \sum_{n=1}^{N}|A_n|(b_n-b_{n+1})\\ &\leqslant K\sum_{n=1}^{N}(b_n-b_{n+1})\\ &= K(b_1-b_{N+1})\\ &< Kb_1.\end{aligned}$$

The result follows from Theorem 3.

THEOREM 30 (*Abel's Test*). *If b_n is a positive, monotonic decreasing function and if Σa_n is convergent, then $\Sigma a_n b_n$ is also convergent.*

Write $c_n = a_n b_n$, $C_n = \sum_{\nu=1}^{n} c_\nu$. Then

$$\begin{aligned} C_n &= a_1 b_1 + a_2 b_2 + \ldots + a_n b_n \\ &= A_1 b_1 + (A_2 - A_1) b_2 + \ldots + (A_n - A_{n-1}) b_n \\ &= A_1(b_1 - b_2) + A_2(b_2 - b_3) + \ldots + A_{n-1}(b_{n-1} - b_n) + A_n b_n, \end{aligned}$$

so that

$$C_n - A_n b_n = \sum_{\nu=1}^{n-1} A_\nu (b_\nu - b_{\nu+1}) \quad \ldots\ldots (1).$$

Since Σa_n is convergent A_n tends to a finite limit. The conditions of the lemma are therefore satisfied so that the series on the right of (1) is convergent. Moreover, by Theorem 3, b_n tends to a finite limit. Hence C_n tends to a finite limit; that is, the series $\Sigma a_n b_n$ is convergent.

THEOREM 31 (*Dirichlet's Test*). *If b_n is a positive, monotonic decreasing function with limit zero, and if, for the series Σa_n, A_n is bounded, then the series $\Sigma a_n b_n$ is convergent.*

Using the notation of Theorem 30 we see from the lemma and relation (1) that $\lim (C_n - A_n b_n)$ is again finite. Now $\lim A_n b_n = 0$ since A_n is bounded and $b_n \to 0$. Thus $\lim C_n$ is finite and the series $\Sigma a_n b_n$ is convergent.

The case $a_n = (-1)^{n-1}$ of Theorem 31 is of considerable importance. We obtain

THEOREM 32. *If b_n is positive and monotonic decreasing with limit zero, then the series $b_1 - b_2 + b_3 - \ldots$ is convergent.*

In other words, in the case of a series whose terms alternate in sign and steadily diminish in magnitude, a necessary and sufficient condition for convergence is that its nth term should tend to zero. For example, the series $\Sigma(-1)^n n^{-a}$, $\Sigma(-1)^n \{\log (n+1)\}^{-a}$ are convergent for $a > 0$, and divergent for $a \leqslant 0$.

Example. Examine for convergence the series

$$\Sigma n^{-a} \sin n\theta,\ \Sigma n^{-a} \cos n\theta.$$

For the series $\Sigma \sin n\theta$, where θ is neither zero nor a multiple of 2π, we have proved (see Art. 19) that A_n is a bounded function of n. By Theorem 31, therefore, the series $\Sigma n^{-a} \sin n\theta$ and, in a similar way, the series $\Sigma n^{-a} \cos n\theta$, are convergent for $a>0$ and for all values of θ except zero or a multiple of 2π. For such values of θ both series diverge for $a \leqslant 0$ since their nth terms do not tend to zero. When θ is a multiple of 2π the first series is a series of zeros and so converges for every value of a. The second, however, reduces to Σn^{-a} which is only convergent when $a>1$.

38. Riemann's Theorem. This theorem, though not of practical importance, is of considerable theoretical interest.

THEOREM 33. *By an appropriate rearrangement of the terms of a conditionally convergent series Σa_n we can make it converge to any given number σ.*

Write

$$b_n = a_n\,,\ (a_n \geqslant 0)\,,\ b_n = 0\,,\ (a_n < 0),$$
$$c_n = a_n\,,\ (a_n \leqslant 0)\,,\ c_n = 0\,,\ (a_n > 0).$$

Then

$$a_n = b_n + c_n\,,\ |a_n| = b_n - c_n.$$

Let $A_n^* = \sum_{r=1}^{n} |a_r|$. Then, with our usual notation,

$$B_n = \tfrac{1}{2}(A_n + A_n^*)\,,\ C_n = \tfrac{1}{2}(A_n - A_n^*).$$

Now A_n tends to a finite limit and A_n^* tends to infinity, so that Σb_n is a properly divergent series of non-negative terms and Σc_n is a properly divergent series of non-positive terms.

We now form a new series Σu_n in the following way. Let n_1 be the least integer such that $\sum_{r=1}^{n_1} b_r > \sigma$ and define u_r to be b_r for $r = 1, 2, \ldots, n_1$. Let n_2 be the least integer such that $\sum_{r=1}^{n_1} b_r + \sum_{r=1}^{n_2} c_r < \sigma$, and define u_{n_1+r} to be c_r for $r = 1$, $2, \ldots, n_2$. Now take n_3 terms of the series Σb_n, where n_3 is just large enough to make $\sum_{r=1}^{n_1+n_3} b_r + \sum_{r=1}^{n_2} c_r > \sigma$, and define $u_{n_1+n_2+r}$ to be b_r for $r = n_1+1, n_1+2, \ldots, n_1+n_3$, and so on. If $U_n = \sum_{r=1}^{n} u_r$ we see that

$$U_{n_1} > \sigma,\ U_{n_1+n_2} < \sigma,\ U_{n_1+n_2+n_3} > \sigma,\ \ldots$$

and that

$$|U_{n_1} - \sigma| < |u_{n_1}|,\ |U_{n_1+n_2} - \sigma| < |u_{n_1+n_2}|,\ \ldots$$

When n lies between n_1 and n_1+n_2, $U_n - \sigma$ lies between $U_{n_1+n_2} - \sigma$ and $U_{n_1} - \sigma$ so that $|U_n - \sigma|$ is not greater than $|u_{n_1}| + |u_{n_1+n_2}|$.

Since the series Σa_n is convergent, given ϵ, we can find $N = N(\epsilon)$ such that $|a_n| < \frac{1}{2}\epsilon$ whenever $n > N$. Let n be any integer greater than both N and n_1. Then we can find an integer $l (\geqslant 1)$ such that

$$n_1 + n_2 + \ldots + n_l \leqslant n < n_1 + n_2 + \ldots + n_l + n_{l+1}$$

and

$$\begin{aligned}|U_n - \sigma| &< |u_{n_1+n_2+\ldots+n_l}| + |u_{n_1+n_2+\ldots+n_{l+1}}| \\ &< \epsilon.\end{aligned}$$

Hence the series Σu_n converges to the sum σ. The series Σu_n contains, besides the terms of the series Σa_n, an infinite number of zero terms. It is clear, however, that the series Σv_n, which is obtained from Σu_n by omitting those zeros which do not occur in the original series Σa_n, is also convergent to the sum σ.

The above proof may be modified to show that, by a suitable rearrangement of the terms, a conditionally convergent series may be made to be properly divergent or to oscillate finitely or infinitely.

39. Complex Limits. Let U_n be a complex function of the real variable n. Then we say that U_n tends to the limit u as n tends to infinity if, given ϵ, we can find $N = N(\epsilon)$ such that * $|U_n - u| < \epsilon$ whenever $n > N$. Suppose that $U_n = A_n + iB_n$ and that $u = \alpha + i\beta$.

Then $$U_n - u = A_n - \alpha + i(B_n - \beta)$$

and $$\left.\begin{matrix}|A_n - \alpha| \\ |B_n - \beta|\end{matrix}\right\} \leqslant |U_n - u| \leqslant |A_n - \alpha| + |B_n - \beta|.$$

It follows that, to say that $U_n = A_n + iB_n$ tends to the limit $\alpha + i\beta$ is the same as saying that $A_n \to \alpha$ and $B_n \to \beta$.

It is easy to see that the proofs of the fundamental limit theorems can be modified so as to apply to the case of complex functions. Moreover, Theorem 6 remains true, the proof of the necessity of the condition being as before. For the sufficiency of the condition we note that, if we write $U_n = A_n + iB_n$, then $|A_{n+p} - A_n|$ and $|B_{n+p} - B_n|$, being each less than $|U_{n+p} - U_n|$, are less than ϵ whenever $n > N$ and for all values of p. It follows that A_n and B_n each tend to finite limits and hence that U_n tends to a definite limit.

40. Series whose Terms may be Complex. Let Σu_n be a series, some or all of whose terms are complex. If $U_n = \sum_{r=1}^{n} u_r$, then the series is said to converge or diverge according as U_n tends to a definite limit or not. The series is said to be absolutely convergent if $\Sigma |u_n|$ is convergent.

Let $u_n = a_n + ib_n$. The preceding article shows that to discuss the convergence of the series Σu_n is the same as discussing the convergence of the two real series Σa_n and Σb_n. Thus all the theorems which we have proved for real series have straightforward analogues in the case of complex series. In particular, an absolutely convergent complex series may have its terms rearranged without

* If $z = x + iy$ is a complex number, then $|z| = (x^2 + y^2)^{\frac{1}{2}}$. If z_1, z_2 are any two complex numbers, then $|z_1 + z_2| \leqslant |z_1| + |z_2|$ and, whati s in reality the same inequality, $|z_1 \pm z_2| \geqslant |z_1| - |z_2|$. *Cf.* Phillips, *Functions of a Complex Variable*, § 2.

affecting its convergence or its sum. Since complex series do not differ materially from a pair of real series, we shall assume throughout the remainder of the book that, unless otherwise stated, all the series with which we deal are real series.

41. Abel's Lemma. We conclude this chapter by proving a theorem of considerable importance.

THEOREM 34. *If b_n is a positive monotonic decreasing sequence and if $h(m, n)$, $H(m, n)$ denote respectively the least and greatest values of the sums $\sum_{r=m}^{\nu} a_r$ for $\nu = m$, $m+1, \ldots, n$, then*

$$b_m h(m, n) \leqslant \sum_{r=m}^{n} a_r b_r \leqslant b_m H(m, n).$$

Let $f(m, \nu) = \sum_{r=m}^{\nu} a_r$. Then

$$\begin{aligned}\sum_{r=m}^{n} a_r b_r &= a_m b_m + a_{m+1} b_{m+1} + \ldots + a_n b_n \\ &= f(m, m) b_m + \{f(m, m+1) - f(m, m)\} b_{m+1} + \ldots \\ &\qquad + \{f(m, n) - f(m, n-1)\} b_n \\ &= f(m, m)(b_m - b_{m+1}) + f(m, m+1)(b_{m+1} - b_{m+2}) + \ldots \\ &\qquad + f(m, n-1)(b_{n-1} - b_n) + f(m, n) b_n.\end{aligned}$$

Now $b_m - b_{m+1}$, $b_{m+1} - b_{m+2}$, ... are all non-negative. Hence

$$\begin{aligned}&h(m, n)(b_m - b_{m+1} + b_{m+1} - b_{m+2} + \ldots - b_n + b_n) \\ &\qquad \leqslant \sum_{r=m}^{n} a_r b_r \leqslant H(m, n)(b_m - b_{m+1} + \ldots - b_n + b_n)\end{aligned}$$

from which the result follows.

The enunciation of the theorem may be modified in the following way in order to cover the case when a_n may be complex.

If b_n is a positive monotonic decreasing sequence and if $K(m, n)$ denotes the largest of the sums $|\sum_{r=m}^{\nu} a_r|$ for $\nu = m, m+1, \ldots n$, then

$$|\sum_{r=m}^{\nu} a_r b_r| \leqslant b_m K(m, n).$$

Example. Show that, for each fixed value of θ which is not zero or a multiple of 2π,

$$\frac{\cos n\theta}{\log n} + \frac{\cos (n+1)\theta}{\log (n+1)} + \ldots + \frac{\cos 2n\theta}{\log 2n} = O\left(\frac{1}{\log n}\right).$$

By Abel's lemma the absolute value of the left-hand side is not greater than

$$\frac{1}{\log n} K(n, 2n),$$

where $K(n, 2n)$ is the largest of the sums

$$|\cos n\theta + \cos (n+1)\theta + \ldots + \cos (n+\nu)\theta|$$

for $\nu = 0, 1, 2, \ldots n$. Clearly $K(n, 2n) \leqslant 1/|\sin \frac{1}{2}\theta|$ so that the result follows.

Examples

1. If a_n and b_n are real and if the series Σa_n^2, Σb_n^2 are convergent, show that the series $\Sigma a_n b_n$ is absolutely convergent.

2. Determine for what values of x each of the following series is (*a*) absolutely convergent, (*b*) convergent :—

(i) $\Sigma \dfrac{n+3}{(n+1)(n+2)} x^n$, (ii) $\Sigma \dfrac{n!}{(2n)!} x^n$, (iii) $\Sigma \dfrac{(n!)^2}{(2n)!} x^n$,

(iv) $\Sigma \dfrac{(2x-1)^n}{\sqrt{n}}$, (v) $\Sigma (-1)^n \dfrac{1}{(nx)^n}$, (vi) $\Sigma \dfrac{x^{n+1}}{x^n - 1}$,

(vii) $\Sigma \dfrac{x^n}{\sqrt{n}} \log \dfrac{2n+1}{n}$, (viii) $\Sigma \dfrac{x^n}{n^x}$, (ix) $\Sigma (\log x)^n \log \dfrac{n+1}{n}$.

3. For what values of x are the following series convergent ?

(i) $\Sigma \dfrac{\cos nx}{\sqrt{n}}$, (ii) $\Sigma \dfrac{\sin nx}{\log n}$, (iii) $\Sigma \cos nx \sin \dfrac{x}{n}$.

4. Prove that the series

$$1+\tfrac{1}{2}-\tfrac{1}{3}+\tfrac{1}{4}+\tfrac{1}{5}-\tfrac{1}{6}+\tfrac{1}{7}+\tfrac{1}{8}-\tfrac{1}{9}+\ldots$$

is divergent, while the series

$$1+\tfrac{1}{2}-\tfrac{2}{3}+\tfrac{1}{4}+\tfrac{1}{5}-\tfrac{2}{6}+\tfrac{1}{7}+\tfrac{1}{8}-\tfrac{2}{9}+\ldots$$

converges to the sum log 3.

5. Discuss the convergence of the series

$$\sum_{n=0}^{\infty} \frac{(3n+4)}{n(n+1)(n+2)}\, x^n,$$

and find its sum when $x = 1$.

6. Prove that the series

$$1+\frac{1}{3^\lambda}-\frac{1}{2^\lambda}+\frac{1}{5^\lambda}+\frac{1}{7^\lambda}-\frac{1}{4^\lambda}+\ldots+\frac{1}{(4n-3)^\lambda}$$
$$+\frac{1}{(4n-1)^\lambda}-\frac{1}{(2n)^\lambda}+\ldots$$

is properly divergent for $\lambda<1$ and convergent for $\lambda \geqslant 1$. Show that when $\lambda = 1$ the sum of the series is $\log(2\sqrt{2})$.

ANSWERS. 2. (i) (*a*) $|x|<1$, (*b*) $-1\leqslant x<1$; (ii) (*a*), (*b*) all values of x; (iii) (*a*) $|x|<4$, (*b*) $-4\leqslant x<4$; (iv) (*a*) $0<x<1$, (*b*) $0\leqslant x<1$; (v) (*a*), (*b*) $x\neq 0$; (vi) (*a*), (*b*) $|x|<1$; (vii) (*a*) $|x|<1$, (*b*) $-1\leqslant x<1$; (viii) (*a*), (*b*) $|x|<1$; (ix) (*a*) $1/e<x<e$, (*b*) $1/e\leqslant x<e$. 3. (i) $x\neq 2k\pi$, where $k=0, \pm 1, \pm 2, \ldots$; (ii) all values of x; (iii) $x\neq 2k\pi$, where $k=\pm 1, \pm 2, \ldots$. 5. Series converges absolutely if $|x|\leqslant 1$, $2\tfrac{1}{2}$.

CHAPTER VI

SERIES OF FUNCTIONS

42. Uniform Convergence. Suppose that $A_n(x)$ is a function of the integral variable n and of the continuous variable x which is defined for all positive integral values of n and for all values of x in the interval $a \leqslant x \leqslant b$. Suppose further that, for each value of x in the interval (a, b), the function $A_n(x)$ tends to a definite limit as $n \to \infty$. This limit will be a function of x which we shall denote by $\alpha(x)$. From the definition of a limit it follows that, given ϵ, we can determine a positive integer N such that $|\alpha(x) - A_n(x)| < \epsilon$ whenever $n > N$. As a rule this integer N, besides depending on ϵ, will also depend on x. If, however, it is possible, for any given ϵ, to determine an integer N, *which is independent of* x, such that $|\alpha(x) - A_n(x)| < \epsilon$ whenever $n > N$, then we say that, as $n \to \infty$, the function $A_n(x)$ **tends uniformly** or **converges uniformly** to $\alpha(x)$ for $a \leqslant x \leqslant b$.

To illustrate these points consider the function

$$A_n(x) = x^n, \ (0 \leqslant x \leqslant \tfrac{1}{2}).$$

Given $\epsilon(<1)$ we have $|A_n(x)| < \epsilon$ if $x^n < \epsilon$; that is, if $n > (\log\epsilon/\log x)$. Hence if we take N to be $[\log\epsilon/\log x]$, we shall have $|A_n(x)| < \epsilon$ whenever $n > N$. Of all the values of N corresponding to the various values of x the largest is $[\log\epsilon/\log\frac{1}{2}]$. Thus, for all values of x in the interval $(0, \frac{1}{2})$ we can write $|A_n(x)| < \epsilon$ whenever $n > [\log\epsilon/\log\frac{1}{2}]$. We therefore conclude that, as $n \to \infty$, $A_n(x)$ tends uniformly to zero.

Again, if

$$A_n(x) = x^n, \ (0 \leqslant x < 1),$$
$$= 0, \ (x = 1),$$

then $|A_n(x)| < \epsilon$ if $n > [\log\epsilon/\log x]$. In this case the function $[\log\epsilon/\log x]$ has no largest value for the various values of x under consideration and, although $A_n(x)$ tends to zero for each value of x, it does not tend uniformly to zero.

In the light of these examples we may therefore rewrite our definition of uniform convergence as follows. If, for a certain range of values of x, given ϵ, we can find $N = N(\epsilon, x)$ such that $|a(x) - A_n(x)| < \epsilon$ whenever $n > N$, then $A_n(x)$ converges uniformly to $a(x)$ if $N(\epsilon, x)$ is a bounded function of x.

Although we have defined uniform convergence with reference to a finite closed interval $a \leqslant x \leqslant b$ it is clearly unnecessary for x to be so restricted. The definition remains essentially unaltered for intervals such as $a < x < b$, $x \geqslant a$, etc. It also applies to cases when x may take any infinite set of values. For example, we may speak of the function $A_n(M)$ converging uniformly for all positive integral values of M.

43. Series of Functions. Let $a_n(x)$ be a function of n and x defined for all positive integral values of n and for $a \leqslant x \leqslant b$ and let $A_n(x) = \sum_{r=1}^{n} a_r(x)$. The series $\Sigma a_n(x)$, if convergent, will have a sum $a(x) = \lim_{n\to\infty} A_n(x)$, which will necessarily be a function of x. The series is said to **converge uniformly** to the sum $a(x)$ for $a \leqslant x \leqslant b$ if $A_n(x)$ tends uniformly to $a(x)$ for $a \leqslant x \leqslant b$.

The fundamental theorem for the uniform convergence of a series may be stated as follows :

THEOREM 35. *A necessary and sufficient condition for the series $\Sigma a_n(x)$ to be uniformly convergent for $a \leqslant x \leqslant b$ is* *that, given* ϵ, *we can find* $N=N(\epsilon)$ *such that* $|\sum_{r=n+1}^{n+p} a_r(x)|<\epsilon$ *whenever* $n>N$ *and for any positive integral value of* p.

The condition is necessary for, if $\Sigma a_n(x)$ is uniformly convergent for $a\leqslant x\leqslant b$, there is a function $\alpha(x)$ with the property that, given ϵ, we can find $N=N(\epsilon)$ such that, for $n>N$ and all values of x in (a, b),

$$|\alpha(x)-A_n(x)|<\tfrac{1}{2}\epsilon.$$

It follows that, for such a value of n and any positive integral value of p,

$$|\alpha(x)-A_{n+p}(x)|<\tfrac{1}{2}\epsilon.$$

Thus, for $a\leqslant x\leqslant b$, $n>N$ and any positive integral value of p,

$$\begin{aligned}|\sum_{r=n+1}^{n+p} a_r(x)| &= |A_{n+p}(x)-A_n(x)| \\ &\leqslant |A_{n+p}(x)-\alpha(x)|+|\alpha(x)-A_n(x)| \\ &< \epsilon.\end{aligned}$$

To show that the condition is sufficient we observe that, if the condition is satisfied, we have, for $a\leqslant x\leqslant b$,

$$A_n(x)-\epsilon<A_{n+p}(x)<A_n(x)+\epsilon,$$

where n is any fixed integer greater than N. Now, for each value of x, the series $\Sigma a_n(x)$ is convergent by Theorem 7; that is, $A_{n+p}(x)$ tends to a definite limit $\alpha(x)$, say, as p tends to infinity. We then have, for $n>N$ and $a\leqslant x\leqslant b$,

$$A_n(x)-\epsilon\leqslant\alpha(x)\leqslant A_n(x)+\epsilon\,;$$

that is, the series $\Sigma a_n(x)$ converges uniformly to the sum $\alpha(x)$.

44. Tests for Uniform Convergence. We now obtain some simple tests for the uniform convergence of series.

Theorem 36. (*Weierstrass's M-test.*) *If, for* $a\leqslant x\leqslant b$ *we have* $|a_n(x)|\leqslant M_n$, *where the series* ΣM_n *is convergent, then the series* $\Sigma a_n(x)$ *is uniformly convergent for* $a\leqslant x\leqslant b$.

Since the series ΣM_n is convergent, given ϵ, we can find $N = N(\epsilon)$ such that, for $n>N$ and any positive integral value of p,

$$M_{n+1}+M_{n+2}+\ldots+M_{n+p}<\epsilon.$$

For all such values of n and p, and for $a \leqslant x \leqslant b$,

$$\left|\sum_{r=n+1}^{n+p} a_r(x)\right| \leqslant \sum_{r=n+1}^{n+p} |a_r(x)| \leqslant \sum_{r=n+1}^{n+p} M_n < \epsilon.$$

The uniform convergence of the series $\Sigma a_n(x)$ for $a \leqslant x \leqslant b$ then follows from Theorem 35.

It is easy to see that the same proof would hold if M_n were a function of x and if the series $\Sigma M_n(x)$ were uniformly convergent for $a \leqslant x \leqslant b$.

For example, the series $\Sigma \dfrac{x^n}{n^2}$ is uniformly convergent for $-1 \leqslant x \leqslant 1$ since, for such values of x,

$$\left|\frac{x^n}{n^2}\right| \leqslant \frac{1}{n^2}$$

and $\Sigma 1/n^2$ is convergent.

Other and more delicate tests for uniform convergence are obtained by making modifications in Theorems 30 and 31.

THEOREM 37. *If $b_n(x)$ is a positive, monotonic decreasing function of n for each value of x in the interval $a \leqslant x \leqslant b$, if $b_n(x)$ is bounded for all values of n and x concerned, and if the series $\Sigma a_n(x)$ is uniformly convergent for $a \leqslant x \leqslant b$, then so also is the series $\Sigma a_n(x) b_n(x)$.*

Suppose that $b_n(x)<K$ for $a \leqslant x \leqslant b$ and all positive integral values of n, where K is independent of x and n. Given ϵ, we can find $N = N(\epsilon)$ such that, for $n>N$ and any positive integral value of ν,

$$\left|\sum_{r=n+1}^{n+\nu} a_r(x)\right| < \epsilon/K.$$

By Theorem 34 we then obtain

$$\left|\sum_{r=n+1}^{n+p} a_r(x)b_r(x)\right| \leqslant b_n(x) \operatorname*{Max}_{\nu=1,2,\ldots p} \left|\sum_{r=n+1}^{n+\nu} a_r(x)\right|$$
$$< K \,.\, \epsilon/K = \epsilon.$$

The theorem therefore follows.

For example, the series $\Sigma \dfrac{(-1)^{n-1}}{n} |x|^n$ is uniformly convergent for $-1 \leqslant x \leqslant 1$, since $|x|^n$ is positive, monotonic decreasing and bounded for $-1 \leqslant x \leqslant 1$ and the series $\Sigma(-1)^{n-1}/n$ is convergent.

THEOREM 38. *If $b_n(x)$ is a positive, monotonic decreasing function of n for each value of x in the range $a \leqslant x \leqslant b$, if $b_n(x)$ tends uniformly to zero for $a \leqslant x \leqslant b$ and if there is a number K, independent of x and n, such that, for all integral values of n and all values of x in (a, b),*

$$\left|\sum_{r=1}^{n} a_r(x)\right| < K,$$

then the series $\Sigma a_n(x)b_n(x)$ is uniformly convergent for $a \leqslant x \leqslant b$.

Given ϵ, we can find $N = N(\epsilon)$ such that, for $n > N$ and all values of x in the range $a \leqslant x \leqslant b$,

$$0 \leqslant b_n(x) < \epsilon/2K.$$

For such a value of n and any positive integral value of p we have, by Abel's Lemma,

$$\left|\sum_{r=n+1}^{n+p} a_r(x)b_r(x)\right| \leqslant b_n(x) \operatorname*{Max}_{\nu=1,2,\ldots p} \left|\sum_{r=n+1}^{n+\nu} a_r(x)\right|$$
$$\leqslant b_n(x)\left\{\left|\sum_{r=1}^{n} a_r(x)\right| + \operatorname*{Max}_{\nu=1,2,\ldots p} \left|\sum_{r=1}^{n+\nu} a_r(x)\right|\right\}$$
$$< \frac{\epsilon}{2K}(K+K)$$
$$= \epsilon.$$

This proves the theorem.

For example, the series $\Sigma\{\log(n+1)\}^{-x}\cos nx$ is uniformly convergent for $0<\theta_1\leqslant x\leqslant\theta_2<2\pi$. When x lies in this range $\{\log(n+1)\}^{-x}$ is a positive monotonic decreasing function of n. Also, since $\{\log(n+1)\}^{-x}\leqslant\{\log(n+1)\}^{-\theta_1}$ the function $\{\log(n+1)\}^{-x}$ tends uniformly to zero as $n\to\infty$. Moreover, in this range,

$$\left|\sum_{r=1}^{n}\cos rx\right|\leqslant 1/(\sin\tfrac{1}{2}x),$$

which in turn is less than or equal to the larger of $1/(\sin\frac{1}{2}\theta_1)$, $1/(\sin\frac{1}{2}\theta_2)$, both of which are independent of x and n. The series is therefore uniformly convergent in the range stated. It is of course to be understood that θ_1 may be as close to zero and θ_2 as close to 2π as we please.

45. Some Properties of Uniformly Convergent Series. We turn now to a consideration of the more important properties of uniformly convergent series.

THEOREM 39. *If the series* $\sum_{n=1}^{\infty}a_n(x)$ *is uniformly convergent for* $a\leqslant x\leqslant b$ *to the sum* $\alpha(x)$ *and if, for each value of* n, $a_n(x)$ *tends to a limit* s_n *as* $x\to x_0$, *where* x_0 *is some point in the range* (a, b), *then, as* $x\to x_0$, $\alpha(x)$ *tends to the limit* σ, *where* σ *is the sum of the series* $\sum_{n=1}^{\infty}s_n$.

In the first place, we observe that the series Σs_n is convergent, for, given ϵ_1, we can find $N=N(\epsilon_1)$ such that, for $n>N$ and any positive integral value of p,

$$-\epsilon_1<a_{n+1}(x)+a_{n+2}(x)+\ldots+a_{n+p}(x)<\epsilon_1.$$

Let $x\to x_0$. Then, for $n>N$ and any positive integral value of p,

$$-\epsilon_1\leqslant s_{n+1}+s_{n+2}+\ldots+s_{n+p}\leqslant\epsilon_1.$$

The convergence of Σs_n then follows from Theorem 6.

Write $S_n = \sum_{r=1}^{n} s_r$. Then, given ϵ_2, we can find positive integers $N_1 = N_1(\epsilon_2)$ and $N_2 = N_2(\epsilon_2)$ such that

$$|\sigma - S_n| < \tfrac{1}{3}\epsilon_2,$$

whenever $n > N_1$, and

$$|\alpha(x) - A_n(x)| < \tfrac{1}{3}\epsilon_2$$

whenever $n > N_2$ and for all values of x in (a, b). Let N be a fixed integer greater than both N_1 and N_2. Then we can determine $\eta = \eta(N, \epsilon_2) = \eta(\epsilon_2)$ such that, for all values of x in (a, b) satisfying the inequality $|x - x_0| < \eta$,

$$|A_N(x) - S_N| < \tfrac{1}{3}\epsilon_2.$$

Collecting these results we have

$$\begin{aligned}|\alpha(x) - \sigma| &\leqslant |\alpha(x) - A_N(x)| + |A_N(x) - S_N| + |S_N - \sigma| \\ &< \tfrac{1}{3}\epsilon_2 + \tfrac{1}{3}\epsilon_2 + \tfrac{1}{3}\epsilon_2 = \epsilon_2\end{aligned}$$

for all values of x in (a, b) satisfying the inequality $|x - x_0| < \eta$. Hence $\alpha(x)$ tends to σ as $x \to x_0$. Clearly the same result is also true if $\Sigma a_n(x)$ is uniformly convergent for $x \geqslant a$ and if $a_n(x)$ tends to s_n as x tends to *infinity*.

It will be observed that, in effect, this theorem is an extension to the case of infinite series of Theorem 1 (i).

THEOREM 40. *If the series $\sum_{n=1}^{\infty} a_n(x)$ is uniformly convergent for $a \leqslant x \leqslant b$ to the sum $\alpha(x)$ and if, at a point x_0 in the range (a, b), each of the functions $a_n(x)$ is continuous, then $\alpha(x)$ is continuous at the point x_0.*

Since $a_n(x)$ is continuous at the point x_0 we have

$$\lim_{x \to x_0} a_n(x) = a_n(x_0).$$

Hence, by Theorem 39,

$$\lim_{x \to x_0} \alpha(x) = \alpha(x_0);$$

that is, $\alpha(x)$ is continuous at the point x_0.

In the case of series whose sum can be readily calculated this theorem often provides a good negative test for uniform convergence. For example, when $a_n(x) = x^n(1-x)$, $0 \leqslant x \leqslant 1$, we have $a(x) = 0$ for $x = 1$, while $a(x) = 1$ for $0 \leqslant x < 1$. For each value of n, $a_n(x)$ is continuous for $0 \leqslant x \leqslant 1$, whereas $a(x)$ is not continuous throughout this range. It follows that the series cannot be uniformly convergent for $0 \leqslant x \leqslant 1$.

THEOREM 41. (*Term by term integration.*) *If the series* $\sum_{n=1}^{\infty} a_n(x)$ *converges uniformly for* $a \leqslant x \leqslant b$ *to the sum* $a(x)$ *and if, for each value of* n, $a_n(x)$ *is continuous in this interval, then the series* $\sum_{n=1}^{\infty} \int_a^x a_n(t)dt$ *converges uniformly for* $a \leqslant x \leqslant b$ *to the sum* $\int_a^x a(t)dt$

In the first place, it should be observed that all the integrals do in fact exist,* since all the functions concerned are continuous.

By hypothesis, given ϵ, we can find $N = N(\epsilon)$ such that, whenever $n > N$,

$$|a(x)-A_n(x)| < \epsilon/(b-a).$$

For such values of n we then have

$$\begin{aligned}\left|\int_a^x \{a(t)-A_n(t)\}dt\right| &\leqslant \int_a^x |a(t)-A_n(t)|dt \\ &< \int_a^x \frac{\epsilon}{b-a}dt \\ &\leqslant \epsilon.\end{aligned}$$

The theorem is therefore proved.

Example. By expanding $1/(1+\cos\theta \cos x)$ in ascending powers of $\cos\theta \cos x$ prove that, for $0 < \theta < \pi$,

$$\operatorname{cosec} \theta = 1 + \sum_{\nu=1}^{\infty} \frac{(2\nu)!}{2^{2\nu}(\nu!)^2} \cos^{2\nu}\theta.$$

---

* See *G.*, p. 74.

For $0<\theta<\pi$ and all values of x we have

$$\frac{1}{1+\cos\theta\cos x} = 1+\sum_{n=1}^{\infty}(-1)^n \cos^n\theta \cos^n x.$$

The series on the right is uniformly convergent for all values of x since

$$|(-1)^n \cos^n\theta \cos^n x| \leqslant |\cos\theta|^n,$$

and $\Sigma\ |\cos\theta|^n$ is convergent. Hence

$$\int_0^\pi \frac{dx}{1+\cos\theta\cos x} = \pi+\sum_{n=1}^{\infty}(-1)^n \cos^n\theta \int_0^\pi \cos^n x dx.$$

Putting $t = \tan\frac{1}{2}x$ this becomes

$$\int_0^\infty \frac{2dt}{t^2(1-\cos\theta)+(1+\cos\theta)} = \pi+2\sum_{\nu=1}^{\infty}\cos^{2\nu}\theta \int_0^{\frac{1}{2}\pi}\cos^{2\nu}x dx;$$

that is,*

$$\frac{2}{\sin\theta}\left[\tan^{-1}\left\{t\sqrt{\left(\frac{1-\cos\theta}{1+\cos\theta}\right)}\right\}\right]_0^\infty$$

$$= \pi\left\{1+\sum_{\nu=1}^{\infty}\frac{(2\nu-1)(2\nu-3)\ldots 3.1}{2\nu(2\nu-2)\ldots 4.2}\cos^{2\nu}\theta\right\},$$

whence, for $0<\theta<\pi$,

$$\operatorname{cosec}\theta = 1+\sum_{\nu=1}^{\infty}\frac{(2\nu)!}{2^{2\nu}(\nu!)^2}\cos^{2\nu}\theta.$$

THEOREM 42. (*Term by term differentiation.*) *If the series $\Sigma a_n(x)$ converges to the sum $a(x)$ for $a\leqslant x\leqslant b$, if $a_n'(x)$ is continuous for $a\leqslant x\leqslant b$ and if $\Sigma a_n'(x)$ is uniformly convergent for $a\leqslant x\leqslant b$, then the sum of the series $\Sigma a_n'(x)$ is $a'(x)$.*

Suppose that the sum of the series $\Sigma a_n'(x)$ is $\sigma(x)$. By Theorem 41, if x is any point of (a, b),

$$\int_a^x \sigma(t)dt = \sum_{n=1}^{\infty}\int_a^x a_n'(t)dt = \sum_{n=1}^{\infty}\{a_n(x)-a_n(a)\}$$
$$= a(x)-a(a).$$

---

* See *G.*, p. 19.

Since $\sigma(x)$ is continuous for $a \leqslant x \leqslant b$ it follows * that $\alpha(x)$ can be differentiated and that its derivative is $\sigma(x)$.

Example. Show that, for $-1<x<1$,

$$\frac{1}{1+x}+\frac{2x}{1+x^2}+\frac{4x^3}{1+x^4}+\frac{8x^7}{1+x^8}+\ldots=\frac{1}{1-x}.$$

The n-th partial sum of the series

$$\log(1-x)+\log(1+x)+\log(1+x^2)+\log(1+x^4)+\ldots$$

is equal to

$$\begin{aligned} &\log\{(1-x)(1+x)(1+x^2)\ldots(1+x^{2^{n-2}})\} \\ =&\log\{(1-x^2)(1+x^2)\ldots(1+x^{2^{n-2}})\} \\ =&\ldots\ldots \\ =&\log(1-x^{2^{n-1}}) \\ \to& 0 \end{aligned}$$

as $n\to\infty$ for $-1<x<1$. Moreover, for $|x|\leqslant\rho<1$,

$$\left|\frac{2^n x^{2^n-1}}{1+x^{2^n}}\right| \leqslant 2^n \rho^{2^n-1},$$

and the series $\Sigma 2^n \rho^{2^n}$ is convergent. Hence the series

$$\frac{2x}{1+x^2}+\frac{4x^3}{1+x^4}+\frac{8x^7}{1+x^8}+\ldots$$

is uniformly convergent for $|x|\leqslant\rho<1$ and, by Theorem 42, its sum is the derivative of $-\log(1-x)-\log(1+x)$; that is $\frac{1}{1-x}-\frac{1}{1+x}$. The result required follows at once.

Neither Theorem 41 nor Theorem 42 has been stated in its most general form, but what we have obtained is quite general enough to suit most ordinary requirements.

Example. Show that, for $0<\theta<2\pi$,

$$\sum_{\nu=1}^{\infty}\frac{1}{\nu}\cos\nu\theta=-\log(2\sin\tfrac{1}{2}\theta);\quad \sum_{\nu=1}^{\infty}\frac{1}{\nu}\sin\nu\theta=\tfrac{1}{2}(\pi-\theta).$$

---

* See *G.*, p. 79.

Let $z = \cos\theta + i\sin\theta$. Then

$$\sum_{\nu=1}^{n} x^{\nu-1} z^{\nu} = \frac{z\{1-(xz)^n\}}{1-xz},$$

whence we have, for $|x|<1$,

$$\sum_{\nu=1}^{\infty} x^{\nu-1}(\cos\nu\theta + i\sin\nu\theta) = \frac{\cos\theta + i\sin\theta}{1-x\cos\theta - xi\sin\theta}$$
$$= \frac{(\cos\theta - x) + i\sin\theta}{1-2x\cos\theta + x^2},$$

so that

$$\sum_{\nu=1}^{\infty} x^{\nu-1}\cos\nu\theta = \frac{\cos\theta - x}{1-2x\cos\theta+x^2},$$
$$\sum_{\nu=1}^{\infty} x^{\nu-1}\sin\nu\theta = \frac{\sin\theta}{1-2x\cos\theta+x^2}.$$

These series are uniformly convergent for all values of θ and for $|x|\leqslant\rho<1$. Hence, integrating with respect to x, where $0<x<1$, we have

$$\sum_{\nu=1}^{\infty}\frac{x^{\nu}\cos\nu\theta}{\nu} = \int_0^x \frac{\cos\theta - t}{1-2t\cos\theta+t^2}\,dt = -\tfrac{1}{2}\log(1-2x\cos\theta+x^2),$$

$$\sum_{\nu=1}^{\infty}\frac{x^{\nu}\sin\nu\theta}{\nu} = \sin\theta\int_0^x \frac{dt}{1-2t\cos\theta+t^2} = \sin\theta\int_0^x \frac{dt}{(t-\cos\theta)^2+\sin^2\theta}$$
$$= \left[\tan^{-1}\left(\frac{t-\cos\theta}{\sin\theta}\right)\right]_{t=0}^{t=x}$$
$$= \tan^{-1}\left(\frac{x-\cos\theta}{\sin\theta}\right) - \tan^{-1}(-\cot\theta)$$
$$= \begin{cases} \tan^{-1}\left(\dfrac{x-\cos\theta}{\sin\theta}\right) + \left(\dfrac{\pi}{2}-\theta\right), & (0<\theta<\pi), \\[2ex] \tan^{-1}\left(\dfrac{x-\cos\theta}{\sin\theta}\right) + \left(\dfrac{3\pi}{2}-\theta\right), & (\pi<\theta<2\pi). \end{cases}$$

Suppose that θ is neither zero nor a multiple of 2π. Then the series $\sum_{\nu=1}^{\infty}\frac{\cos\nu\theta}{\nu}$ is convergent, and, for $0\leqslant x\leqslant 1$, x^{ν} is positive, monotonic decreasing and bounded. The series $\sum\frac{x^{\nu}\cos\nu\theta}{\nu}$ is therefore uniformly convergent for $0\leqslant x\leqslant 1$.

Let $x \to 1$. Then by Theorem 39 if θ is neither zero nor a multiple of 2π,

$$\sum_{\nu=1}^{\infty} \frac{1}{\nu} \cos \nu\theta = -\tfrac{1}{2} \log (2-2 \cos \theta)$$
$$= -\log (2 \sin \tfrac{1}{2}\theta), \quad (0<\theta<2\pi).$$

In the same way we also obtain

$$\sum_{\nu=1}^{\infty} \frac{\sin \nu\theta}{\nu} = \begin{cases} \tan^{-1} (\tan \tfrac{1}{2} \theta) + \tfrac{1}{2} \pi - \theta, & (0<\theta<\pi), \\ \tan^{-1} (\tan \tfrac{1}{2} \theta) + \tfrac{3}{2} \pi - \theta, & (\pi<\theta<2\pi), \end{cases}$$
$$= \tfrac{1}{2}(\pi-\theta), \quad (0<\theta<2\pi).$$

46. Power Series. The simplest and most important case of a series of functions is the series * $\sum_{n=0}^{\infty} a_n x^n$. Such a series is called a power series. We shall confine ourselves here to a short discussion of power series in the *real* variable x.

THEOREM 43. *If* $\overline{\lim} \sqrt[n]{|a_n|} = 1/R$ *then the series* $\Sigma a_n x^n$ *is convergent for* $|x|<R$ *and divergent for* $|x|>R$.

For

$$\overline{\lim_{n\to\infty}} \sqrt[n]{(|a_n||x|^n)} = |x|/R,$$

whence the series $\Sigma a_n x^n$ is absolutely convergent, and therefore convergent, for $|x|<R$ and divergent for $|x|>R$.

The theorem in effect shows that, to every power series $\Sigma a_n x^n$, there corresponds a unique number R (which may be zero or infinity), such that the series is absolutely convergent whenever $|x|<R$ and divergent whenever $|x|>R$.

If x is replaced by the complex variable z the same proof shows that the series $\Sigma a_n z^n$ is convergent whenever the point z lies within the circle $|z| = R$ and is divergent whenever z is outside this circle. For this reason the number R (which may be zero or infinity) is called the radius of convergence of the power series. For example,

* Unless otherwise stated it is to be assumed that the first term of the power series $\Sigma a_n x^n$ is a_0.

the series $\Sigma n^n x^n$, Σx^n, $\Sigma x^n/n^n$ have respectively radii of convergence equal to 0, 1, ∞.

THEOREM 44. *If its radius of convergence is R the power series $\Sigma a_n x^n$ is uniformly convergent for $|x| \leqslant \rho < R$.*

We have

$$|a_n x^n| \leqslant |a_n| \rho^n$$

and $\Sigma |a_n| \rho^n$ is convergent. Thus, by Theorem 36, the series $\Sigma a_n x^n$ is uniformly convergent for $|x| \leqslant \rho < R$.

We at once conclude that a power series may be integrated term by term so long as the limits of integration lie strictly within the range $(-R, R)$. The radius of convergence of the series $\sum_{n=1}^{\infty} n a_n x^{n-1}$ is $1/\overline{\lim} \sqrt[n]{(n|a_n|)}$, which is equal to R since $\sqrt[n]{n} \to 1$. Thus a power series may also be differentiated term by term at any point x strictly within the range $(-R, R)$.

We now prove an important theorem due to Abel.

THEOREM 45. *If the radius of convergence of the series $\sum_{n=0}^{\infty} a_n x^n$ is R and if $\Sigma a_n R^n$ is convergent, then*

$$\lim_{x \to R} \left(\sum_{n=0}^{\infty} a_n x^n\right) = \sum_{n=0}^{\infty} a_n R^n.$$

The result will follow from Theorem 39 if we show that the series $\Sigma a_n x^n$ is uniformly convergent for $0 \leqslant x \leqslant R$. This in turn follows from Theorem 37, since $\Sigma a_n R^n$ is convergent and $(x/R)^n$ is a positive, monotonic decreasing bounded function of n for $0 \leqslant x \leqslant R$.

The most important case of the theorem occurs when $R = 1$. We then obtain

$$\lim_{x \to 1} \left(\sum_{n=0}^{\infty} a_n x^n\right) = \sum_{n=0}^{\infty} a_n$$

if the series Σa_n is convergent.

As an illustration, consider the series $1 - t^2 + t^4 \ldots$, whose sum for $|t|<1$ is $(1+t^2)^{-1}$. Integrating term by term we have, for $-1<x<1$,

$$\tan^{-1} x = \int_0^x \frac{dt}{1+t^2} = x - \frac{x^3}{3} + \frac{x^5}{5} - \dots$$

Let $x \to 1$. Then

$$\frac{\pi}{4} = 1 - \frac{1}{3} + \frac{1}{5} - \dots$$

which gives Gregory's series for π.

The binomial series

$$1 + \frac{\lambda}{1} x + \frac{\lambda(\lambda-1)}{1.2} x^2 + \dots$$

has unit radius of convergence and its sum, for $-1<x<1$, is $(1+x)^\lambda$. We shall now examine this series in the cases $x = -1$ and $x = 1$.

Writing $-x$ for x we obtain, for $-1<x<1$,

$$(1-x)^\lambda = 1 - \lambda x + \frac{\lambda(\lambda-1)}{1.2} x^2 - \dots .$$

When $x = 1$ the series on the right (see Art. 31) converges for $\lambda \geqslant 0$ and diverges for $\lambda<0$. Its sum for $\lambda>0$ is zero by Theorem 45 and when $\lambda = 0$ its sum is obviously unity. It follows that, when $x = -1$, the original series is convergent to the sum zero for $\lambda>0$, is convergent to the sum 1 when $\lambda = 0$ and is divergent when $\lambda<0$.

When $x = 1$ the binomial series becomes

$$1 + \frac{\lambda}{1} + \frac{\lambda(\lambda-1)}{1.2} + \dots .$$

Denoting it by $\sum_{n=0}^{\infty} a_n$ we have

$$a_n = (-1)^n \frac{(n-\lambda-1)(n-\lambda-2)\dots(1-\lambda)(-\lambda)}{1.2\dots n}.$$

If $\lambda \leqslant -1$. $|a_n| \geqslant 1$ so that the series is divergent. If $\lambda > -1$

the terms of the series ultimately alternate in sign and $|a_n|$ steadily decreases. The convergence of the series therefore depends on whether or not $a_n \to 0$. Writing $\rho = [\lambda]$, and remembering that $1-x<e^{-x}$, $(x>0)$, we have

$$a_n = O\left\{\left(1-\frac{\lambda+1}{n}\right)\left(1-\frac{\lambda+1}{n-1}\right)\cdots\left(1-\frac{1+\lambda}{\rho+2}\right)\right\}$$
$$= O\left\{e^{-(\lambda+1)(1+\frac{1}{2}+\dots+\frac{1}{n})}\right\}$$
$$= O\{e^{-(\lambda+1)\log n}\}$$
$$= o\,(1),$$

when $\lambda > -1$. The series therefore converges when $\lambda > -1$ and its sum, by Theorem 45, is 2^λ.

The example just considered shows that it is not true to assert that if the series $\Sigma a_n x^n$ converges for $-1<x<1$ and if $\lim_{x\to 1} (\Sigma a_n x^n)$ is finite then the series converges at the point $x = 1$. That this converse of Theorem 45 is false follows from the fact that, for all values of λ, $(1+x)^\lambda \to 2^\lambda$ as $x \to 1$, whereas the binomial series at $x = 1$ is only convergent for $\lambda > -1$.

We conclude this chapter by proving a theorem of considerable theoretical interest.

THEOREM 46. *Every power series is the Maclaurin series of its sum function.*

Let $a(x) = \sum_{n=0}^{\infty} a_n x^n$. Then $a(x)$ is defined for all values of x inside the range $(-R, R)$, where R is the radius of convergence of the series. For such values of x we have

$$a^{(r)}(x) = \sum_{n=r}^{\infty} n(n-1)\dots(n-r+1)a_n x^{n-r}$$

so that $a^{(r)}(0) = r!\, a_r$. Thus $a_r = a^{(r)}(0)/r!$ for all positive integral values of r. This proves the theorem.

Examples

1. Discuss the uniform convergence with respect to x of the series

$$\text{(i) } \sum_{n=1}^{\infty} \frac{x^n}{n^n}, \qquad \text{(ii) } \sum_{n=1}^{\infty} \frac{x^n}{n^{\alpha}}, \qquad \text{(iii) } \sum_{n=1}^{\infty} \frac{1}{(x^2+n)(x^2+n+1)},$$

$$\text{(iv) } \sum_{n=1}^{\infty} \frac{(-1)^n}{\sqrt{n}} \sin\left(1+\frac{x}{n}\right), \quad \text{(v) } \sum_{n=0}^{\infty} \frac{\sinh x}{\cosh nx \cosh (n+1)x}.$$

2. Discuss the uniform convergence with respect to θ, where θ lies in the range $(0, 2\pi)$, of the series

$$\text{(i) } \sum_{n=2}^{\infty} \frac{\log n}{n} \sin n\theta, \qquad \text{(ii) } \sum_{n=1}^{\infty} \cos^n\theta \cos n\theta,$$

$$\text{(iii) } \sum_{n=1}^{\infty} \frac{\theta(2\pi-\theta) \sin n\theta}{\sqrt{n}}, \qquad \text{(iv) } \sum_{n=1}^{\infty} n^{-\theta} \cos (2n+1)\theta.$$

3. Find the sums, for $|x| \leqslant 1$, of the series

$$\text{(i) } x+\sum_{n=1}^{\infty} \frac{1.3.5\ldots(2n-1)}{2.4.6\ldots 2n} \frac{x^{2n+1}}{2n+1},$$

$$\text{(ii) } x+\sum_{n=1}^{\infty} (-1)^n \frac{1.3.5\ldots(2n-1)}{2.4.6\ldots 2n} \frac{x^{2n+1}}{2n+1},$$

and deduce that

$$1+\frac{1.3}{2.4}\cdot\frac{1}{5}+\frac{1.3.5.7}{2.4.6.8}\cdot\frac{1}{9}+\ldots = \frac{1}{4}\pi+\frac{1}{2}\log (1+\sqrt{2}).$$

4. Show that the series $\sum_{n=0}^{\infty} x(1-x)^n$ is convergent but not uniformly convergent for $0 \leqslant x \leqslant \rho < 2$. Is there an interval of uniform convergence ? Show that the sum of the series is not continuous at the origin but that term by term integration over the range $(0, 1)$ leads to a correct result.

5. Show that the series $\sum_{n=1}^{\infty} x^n(1-x^n)$ is not uniformly convergent in the interval $0 \leqslant x \leqslant 1$ and determine for what values of α the series $\sum_{n=1}^{\infty} (1-x)^{\alpha} x^n(1-x^n)$ is uniformly convergent in that interval.

6. By comparing it with an integral, show that the series

$$\sum_{n=1}^{\infty} \frac{x^{a}}{1+n^{2}x^{2\beta}}, \quad (a>0,\ \beta>0),$$

will not be uniformly convergent in any interval including the origin if $a \leqslant \beta$.

7. If

$$f(x) = \sum_{n=0}^{\infty} \left\{ \frac{x^{2n+1}}{2n+1} - \frac{x^{2n+2}}{2n+2} \right\}, \quad \phi(x) = \sum_{n=0}^{\infty} \left\{ \frac{x^{2n+1}}{2n+1} - \frac{x^{n+1}}{2n+2} \right\},$$

show that $f(x)$ is continuous for $0 \leqslant x \leqslant 1$ and that $\phi(x)$ is continuous in the same interval except at the point $x = 1$. Explain the discrepancy.

8. Prove that

(i) $\sin\theta - \frac{1}{2}\sin 2\theta + \frac{1}{3}\sin 3\theta - \ldots = \frac{1}{2}\theta, \ (-\pi < \theta < \pi)$,

(ii) $\cos\theta - \frac{1}{2}\cos 2\theta + \frac{1}{3}\cos 3\theta - \ldots = \log(2\cos\frac{1}{2}\theta), \ (-\pi < \theta < \pi)$,

(iii) $\sin\theta + \frac{1}{3}\sin 3\theta + \frac{1}{5}\sin 5\theta + \ldots = \frac{1}{4}\pi, \ (0 < \theta < \pi)$,

(iv) $\dfrac{1}{1^2} + \dfrac{1}{2^2} + \dfrac{1}{3^2} + \ldots = \dfrac{\pi^2}{6}$,

(v) $\dfrac{\cos\theta}{1^2} - \dfrac{\cos 2\theta}{2^2} + \dfrac{\cos 3\theta}{3^2} - \ldots = \dfrac{\pi^2}{12} - \dfrac{\theta^2}{4}, \ (-\pi \leqslant \theta \leqslant \pi)$.

[To obtain (iv), multiply (i) by θ and integrate from 0 to π, justifying the term by term integration over this range.]

9. Prove that, for $0 \leqslant \theta \leqslant 2\pi$,

$$\sum_{n=1}^{\infty} \frac{\cos n\theta}{n(n+1)} = 1 - 2\sin\tfrac{1}{2}\theta\{\sin\tfrac{1}{2}\theta \log(2\sin\tfrac{1}{2}\theta) + \tfrac{1}{2}(\pi-\theta)\cos\tfrac{1}{2}\theta\},$$

$$\sum_{n=1}^{\infty} \frac{\sin n\theta}{n(n+1)} = 2\sin\tfrac{1}{2}\theta\{\tfrac{1}{2}(\pi-\theta)\sin\tfrac{1}{2}\theta - \cos\tfrac{1}{2}\theta \log(2\sin\tfrac{1}{2}\theta)\}.$$

Deduce that

(i) $\dfrac{1}{1.2} - \dfrac{1}{3.4} + \dfrac{1}{5.6} - \ldots = \frac{1}{4}(\pi - 2\log 2)$,

(ii) $\dfrac{1}{2.3} - \dfrac{1}{4.5} + \dfrac{1}{6.7} - \ldots = \frac{1}{4}(\pi + 2\log 2 - 4)$.

10. Prove that, if $-1 \leqslant x \leqslant 1$,

$$\int_0^1 \frac{1-t}{1-xt^3}\,dt = \frac{1}{1.2}+\frac{x}{4.5}+\frac{x^2}{7.8}+\ldots,$$

and deduce that

$$\text{(i)}\quad \frac{1}{1.2}+\frac{1}{4.5}+\frac{1}{7.8}+\ldots = \frac{\pi}{3\sqrt{3}},$$

$$\text{(ii)}\quad \frac{1}{1.2}+\frac{1}{7.8}+\frac{1}{13.14}+\ldots = \frac{\pi}{6\sqrt{3}}+\tfrac{1}{3}\log 2.$$

11. By expanding $(1+\cos\theta\cos x)^{-1}$ in ascending powers of $\cos\theta\cos x$ prove that, for $0<\theta<\pi$,

$$\frac{\theta}{\sin\theta} = \frac{\pi}{2}\sum_{m=0}^{\infty}\frac{(2m)!}{2^{2m}(m!)^2}\cos^{2m}\theta - \sum_{m=1}^{\infty}\frac{2^{2m-2}\{(m-1)!\}^2}{(2m-1)!}\cos^{2m-1}\theta.$$

12. If, for a certain range of values of x, $\sum_{n=0}^{\infty} a_n x^n = \sum_{n=0}^{\infty} b_n x^n$ show that $a_n = b_n$ for all positive integral values of n.

ANSWERS. 1. (i) For all values of x; (ii) $-1 \leqslant x \leqslant 1$ if $a>1$, $-1 \leqslant x \leqslant k<1$ if $0<a \leqslant 1$, $-1<-k \leqslant x \leqslant l<1$ if $a \leqslant 0$; (iii) for all values of x; (iv) for all values of x; (v) for $x \leqslant -k<0$ and for $0<k \leqslant x$. 2. (i) $0<k \leqslant \theta \leqslant l<2\pi$; (ii) $0<k \leqslant \theta \leqslant l<\pi$ and $\pi<p \leqslant \theta \leqslant q<2\pi$; (iii) $0 \leqslant \theta \leqslant 2\pi$; (iv) $0<k \leqslant \theta$. 3. (i) $\sin^{-1}x$; (ii) $\log\{x+\sqrt{(1+x^2)}\}$. 4. $0<k \leqslant x \leqslant \rho<2$. 5. $a>1$. 7. $f(x) = \log(1+x)$, $\phi(x) = \frac{1}{2}\log(1+x)$ when $|x|<1$, while $f(1) = \phi(1) = \log 2$. In the proof of Theorem 45 it is assumed that the series is arranged in ascending powers of x. The theorem does not, therefore, apply to the series for $\phi(x)$ with the brackets removed.

CHAPTER VII

THE MULTIPLICATION OF SERIES

47. Multiplication of Series of Non-Negative Terms. Suppose that $\sum_{n=0}^{\infty} a_n$, $\sum_{n=0}^{\infty} b_n$ are any two series. Then the series $\sum_{n=0}^{\infty} c_n$, where

$$c_n = a_0 b_n + a_1 b_{n-1} + \dots + a_n b_0,$$

is called the product series of the two series Σa_n and Σb_n. The reason for this definition, in the case of series whose terms are non-negative, is shown by the following theorem.

THEOREM 47. *If* $a_n \geqslant 0$, $b_n \geqslant 0$ *and if* $\sum_{n=0}^{\infty} a_n$ *and* $\sum_{n=0}^{\infty} b_n$ *converge respectively to the sums* α *and* β *then* $\sum_{n=0}^{\infty} c_n$ *converges to the sum* $\alpha\beta$.

Consider the array

$$\begin{array}{ccccc}
a_0b_0 & a_1b_0 & a_2b_0 & a_3b_0 & a_4b_0 \;\dots \\
a_0b_1 & a_1b_1 & a_2b_1 & a_3b_1 & \dots \\
a_0b_2 & a_1b_2 & a_2b_2 & a_3b_2 & \dots \\
a_0b_3 & a_1b_3 & a_2b_3 & a_3b_3 & \dots \\
a_0b_4 & & & & \\
\dots & & & & \dots
\end{array}$$

and suppose that d_n denotes the sum of all those terms which belong to the $(n+1)$-th square but not to the n-th square. For example

$$d_0 = a_0 b_0, \quad d_1 = a_0 b_1 + a_1 b_1 + a_1 b_0, \dots$$

Clearly we have *

$$d_0 = A_0B_0,$$
$$d_1 = A_1B_1 - A_0B_0,$$
$$d_2 = A_2B_2 - A_1B_1,$$
$$\dots$$
$$d_n = A_nB_n - A_{n-1}B_{n-1}.$$

Adding we obtain $D_n = A_nB_n$ and, since $A_n \to \alpha$, $B_n \to \beta$, it follows that Σd_n converges to the sum $\alpha\beta$. From Theorems 15 and 14 in turn it then follows that the following series

$$a_0b_0+a_0b_1+a_1b_1+a_1b_0+a_0b_2+a_1b_2+a_2b_2+a_2b_1+a_2b_0+\dots,$$
$$a_0b_0+a_0b_1+a_1b_0+a_0b_2+a_1b_1+a_2b_0+a_0b_3+a_1b_2+a_2b_1+a_3b_0 + \dots,$$
$$a_0b_0+(a_0b_1+a_1b_0)+(a_0b_2+a_1b_1+a_2b_0)+\dots$$

are each convergent to the sum $\alpha\beta$. This proves the theorem.

48. Multiplication of General Series. Consider the series Σa_n and Σb_n where

$$a_0 = a_1 = b_0 = b_1 = 0, \; a_n = b_n = \frac{(-1)^n}{\log n} \; (n \geqslant 2).$$

We have

$$c_n = (-1)^n \left\{ \frac{1}{\log 2 \log (n-2)} + \frac{1}{\log 3 \log (n-3)} + \dots + \frac{1}{\log (n-2) \log 2} \right\},$$

so that, when n is even,

$$c_n \geqslant \frac{n-3}{(\log \frac{1}{2}n)^2} \to \infty,$$

and, when n is odd,

$$c_n \leqslant -\frac{n-3}{\{\log \frac{1}{2}(n-1) \log \frac{1}{2}(n+1)\}} \to -\infty.$$

The series Σc_n therefore does not converge.

* Throughout this chapter A_n denotes $\sum_{\nu=0}^{n} a_\nu$ and B_n, C_n, D_n are defined similarly for the series Σb_n, Σc_n and Σd_n.

This example shows that to ensure the convergence of Σc_n to the product of the sums of the series Σa_n and Σb_n we require, besides the convergence of Σa_n and Σb_n, some further limitation on the behaviour of these series or of Σc_n. One sufficient limitation of this kind, namely $a_n \geqslant 0$, $b_n \geqslant 0$, has already been obtained. The theorems of this article provide further illustrations of this principle.

THEOREM 48. *If Σa_n, Σb_n converge absolutely to the sums α and β, then Σc_n converges absolutely to the sum $\alpha\beta$.*

Since Σa_n and Σb_n are convergent it follows, as in the proof of Theorem 47, that the series

$$a_0b_0+(a_0b_1+a_1b_1+a_1b_0) \\ +(a_0b_2+a_1b_2+a_2b_2+a_2b_1+a_2b_0)+\dots \quad (1)$$

converges to the sum $\alpha\beta$, and, since $\Sigma|a_n|$ and $\Sigma|b_n|$ are convergent, that the series

$$|a_0||b_0|+(|a_0||b_1|+|a_1||b_1|+|a_1||b_0|) \\ +(|a_0||b_2|+|a_1||b_2|+|a_2||b_2|+|a_2||b_1|+|a_2||b_0|)+\dots$$

is convergent. From Theorem 15 it follows that the series

$$a_0b_0+a_0b_1+a_1b_1+a_1b_0+a_0b_2+a_1b_2+a_2b_2+a_2b_1+a_2b_0+\dots (2)$$

is absolutely convergent. Let the sum of this series be σ. Series (1) is obtained from series (2) merely by the insertion of certain brackets. Hence, from Theorem 12, $\sigma = \alpha\beta$. It now follows from Theorem 29 that the series

$$a_0b_0+a_0b_1+a_1b_0+a_0b_2+a_1b_1+a_2b_0+\dots$$

converges to the sum $\alpha\beta$ and, from Theorem 12, that the series Σc_n converges to the sum $\alpha\beta$.

The absolute convergence of Σc_n follows from the fact that

$$\sum_{n=0}^{\infty}|c_n| \leqslant |a_0||b_0|+|a_0||b_1|+|a_1||b_0|+\dots ,$$

which is convergent.

Example. If z is any complex number and $\exp z$ is defined to be $\sum_{n=0}^{\infty} z^n/n!$, prove that

$$\exp z \exp \zeta = \exp (z+\zeta).$$

The series for $\exp z$ and $\exp \zeta$ are absolutely convergent for all values of z and ζ respectively. Hence

$$\begin{aligned}\exp z \exp \zeta &= \sum_{n=0}^{\infty} \left(\frac{z^n}{n!} + \frac{z^{n-1}\zeta}{(n-1)!1!} + \frac{z^{n-2}\zeta^2}{(n-2)!2!} + \ldots + \frac{\zeta^n}{n!}\right) \\ &= \sum_{n=0}^{\infty} \frac{1}{n!}\left\{z^n + \binom{n}{1} z^{n-1}\zeta + \binom{n}{2} z^{n-2}\zeta^2 + \ldots + \zeta^n\right\} \\ &= \sum_{n=0}^{\infty} \frac{1}{n!} (z+\zeta)^n \\ &= \exp (z+\zeta).\end{aligned}$$

THEOREM 49. *If Σa_n, Σb_n converge respectively to the sums α, β and if Σc_n converges then $\sum_{n=0}^{\infty} c_n = \alpha\beta$.*

The series $\sum_{n=0}^{\infty} a_n x^n$, $\sum_{n=0}^{\infty} b_n x^n$, $\sum_{n=0}^{\infty} c_n x^n$ are all absolutely convergent for $-1<x<1$ since, by hypothesis, their radii of convergence are not less than unity. Let their sums be $\alpha(x)$, $\beta(x)$, $\gamma(x)$. The third series is clearly the product series of the first two, so that, by Theorem 48,

$$\gamma(x) = \alpha(x)\beta(x).$$

Let $x \to 1$. Then it follows from Theorem 45 that

$$\sum_{n=0}^{\infty} c_n = \alpha\beta.$$

Example. Prove that

$$\sum_{n=0}^{\infty} (-1)^n \left\{\frac{1}{(n+1)\,.\,1} + \frac{1}{n\,.\,2} + \ldots + \frac{1}{1.(n+1)}\right\} = (\log 2)^2.$$

Write $a_n = b_n = (-1)^n/(n+1)$, $(n \geqslant 0)$. Then Σa_n and Σb_n converge to the sum $\log 2$. Also the product series of Σa_n and Σb_n is Σc_n where

$$c_n = (-1)^n \left\{\frac{1}{(n+1).1} + \frac{1}{n\,.\,2} + \ldots + \frac{1}{1\,.(n+1)}\right\}.$$

When n is even we have, by Theorem 16,

$$c_n < 2\left\{\frac{1}{(n+1).1}+\frac{1}{n\,.\,2}+\ldots+\frac{1}{(\frac{1}{2}n+1)^2}\right\}$$

$$= 2\int_1^{\frac{1}{2}n+1}\frac{dx}{x(n+2-x)}+O\left(\frac{1}{n}\right)$$

$$= \frac{2}{n+2}\int_1^{\frac{1}{2}n+1}\left(\frac{1}{x}+\frac{1}{n+2-x}\right)dx+O\left(\frac{1}{n}\right)$$

$$= \frac{2}{n+2}\left[\log\frac{x}{n+2-x}\right]_1^{\frac{1}{2}n+1}+O\left(\frac{1}{n}\right)$$

$$= \frac{2}{n+2}\log(n+1)+O\left(\frac{1}{n}\right)$$

$$= o(1).$$

Similarly, when n is odd, $c_n = o(1)$.

Moreover, $|c_n|$ is a monotonic decreasing function, for $|c_{n-1}|-|c_n|$

$$= \frac{1}{n\,.\,1}+\frac{1}{(n-1).2}+\ldots+\frac{1}{1\,.\,n}-\frac{1}{(n+1).1}-\frac{1}{n\,.\,2}-\ldots-\frac{1}{1\,.(n+1)}$$

$$= \frac{1}{n}\left(\frac{1}{1}-\frac{1}{2}\right)+\frac{1}{n-1}\left(\frac{1}{2}-\frac{1}{3}\right)+\ldots+\frac{1}{1}\left(\frac{1}{n}-\frac{1}{n+1}\right)-\frac{1}{n+1}$$

$$\geqslant \frac{1}{n}\left\{\frac{1}{1}-\frac{1}{2}+\frac{1}{2}-\frac{1}{3}\ldots+\frac{1}{n}-\frac{1}{n+1}\right\}-\frac{1}{n+1}$$

$$= \frac{1}{n}-\frac{1}{n(n+1)}-\frac{1}{n+1}$$

$$= 0.$$

It follows from Theorem 32 that Σc_n is convergent and, from Theorem 49, that its sum is $(\log 2)^2$.

Finally, we have the following theorem of Mertens.

THEOREM 50. *If Σa_n converges absolutely to the sum α and if Σb_n converges to the sum β, then Σc_n converges to the sum $\alpha\beta$.*

Let the sum of the series $\Sigma|a_n|$ be σ. From hypothesis there is a positive number K such that, for all values of n,

$$|b_1+b_2+\ldots+b_n|<K.$$

Moreover, given ϵ, we can find $N_0 = N_0(\epsilon)$ such that, for $n > N_0$ and any positive integral value of p,

$$|a_{n+1}|+|a_{n+2}|+\dots+|a_{n+p}|<\epsilon/4K,$$

and we can find $N_1 = N_1(\epsilon)$ such that, for $n > N_1$, and any positive integral value of p,

$$|b_n+b_{n+1}+\dots+b_{n+p}|<\epsilon/2\sigma.$$

Let N be any fixed positive integer greater than N_0. Then, taking $n > N+N_1$, we have

$$\begin{aligned}A_nB_n-C_n &= (a_0+a_1+\dots+a_n)(b_0+b_1+\dots+b_n)\\ &\qquad -\{a_0b_0+(a_0b_1+a_1b_0)+(a_0b_2+a_1b_1+a_2b_0)+\dots\\ &\qquad\qquad +(a_0b_n+a_1b_{n-1}+\dots+a_{n-1}b_1+a_nb_0)\}\\ &= a_1b_n+a_2(b_{n-1}+b_n)+a_3(b_{n-2}+b_{n-1}+b_n)+\dots\\ &\qquad +a_n(b_1+b_2+\dots+b_n)\\ &= a_1b_n+a_2(b_{n-1}+b_n)+\dots+a_N(b_{n-N+1}+\dots+b_n)\\ &\qquad +a_{N+1}(b_{n-N}+\dots+b_n)+\dots+a_n(b_1+b_2+\dots+b_n)\\ &= P_n+Q_n,\end{aligned}$$

say. Now

$$|Q_n| \leqslant 2K\{|a_{N+1}|+|a_{N+2}|+\dots+|a_n|\} > \frac{2K\epsilon}{4K} = \tfrac{1}{2}\epsilon.$$

Also,

$$\begin{aligned}|P_n| &\leqslant \{|a_1|+|a_2|+\dots+|a_N|\} \max_{0\leqslant m\leqslant N-1} |b_{n-m}+\dots+b_n|\\ &\leqslant \frac{\sigma\epsilon}{2\sigma} = \tfrac{1}{2}\epsilon.\end{aligned}$$

Thus, for $n > N+N_1$ we have $|A_nB_n-C_n|<\epsilon$; that is, $\lim\,(A_nB_n-C_n) = 0$. In other words, the series Σc_n converges to the sum $\alpha\beta$.

Examples

1. Prove that, for certain values of x and θ which are to be stated,

(i) $\dfrac{1}{1-x}\log\dfrac{1}{1-x} = \sum\limits_{n=1}^{\infty}\left(1+\tfrac{1}{2}+\tfrac{1}{3}+\dots+\dfrac{1}{n}\right)x^n,$

(ii) $2\sum_{n=0}^{\infty} x^n \cos n\theta \sum_{n=0}^{\infty} x^n \sin n\theta = \sum_{n=0}^{\infty} (n+1)x^n \sin n\theta.$

2. Prove that, for certain values of x and θ which are to be stated,

$$\sum_{n=1}^{\infty} \frac{x^n}{n} \cos n\theta \sum_{n=1}^{\infty} \frac{x^n}{n} \sin n\theta = \sum_{n=2}^{\infty} \frac{x^n}{n}\left(1+\tfrac{1}{2}+\ldots+\frac{1}{n-1}\right) \sin n\theta,$$

and deduce from the example at the foot of page 77 that

$$\text{(i)}\ \sum_{n=2}^{\infty}\left(1+\tfrac{1}{2}+\ldots+\frac{1}{n-1}\right)\frac{\sin n\theta}{n} = \tfrac{1}{2}(\theta-\pi)\log\left(2 \sin \tfrac{1}{2}\theta\right),$$

$$(0<\theta<2\pi),$$

$$\text{(ii)}\ \sum_{n=1}^{\infty}(-1)^{n-1}\left(1+\tfrac{1}{2}+\ldots+\frac{1}{2n}\right)\frac{1}{2n+1} = \tfrac{1}{8}\pi \log 2.$$

3. Find partial fractions for

$$\frac{(a+1)(a+3)\ldots(a+2n-1)}{a(a+2)\ldots(a+2n)}$$

and hence prove that, if $a>0$ and $-1 \leqslant x < 1$,

$$\left\{1+\tfrac{1}{2}\frac{a}{a+2}x+\frac{1.3}{2.4}\frac{a}{a+4}x^2+\ldots\right\}\left\{1+\tfrac{1}{2}x+\frac{1.3}{2.4}x^2+\ldots\right\}$$

$$= 1+\frac{a+1}{a+2}x+\frac{(a+1)(a+3)}{(a+2)(a+4)}x^2+\ldots$$

Answers. 1. (i) $|x|<1$; (ii) $|x|<1$, all values of θ. 2. $|x|<1$, all values of θ; or $x=1$, $\theta \neq 2k\pi$; or $x=-1$, $\theta \neq (2k+1)\pi$, k being any integer.

CHAPTER VIII

INFINITE PRODUCTS

49. Convergence and Divergence of Infinite Products. Suppose that a_n is any real function of n defined for all positive integral values of n and let

$$P_n = \prod_{r=1}^{n}(1+a_r) = (1+a_1)(1+a_2)\ldots(1+a_n).$$

If P_n tends to a finite non-zero limit P, then we say that the infinite product $\Pi(1+a_n)$ **converges** to the limit P and we write

$$\prod_{n=1}^{\infty}(1+a_n) = P.$$

If P_n does not tend to a finite non-zero limit, then we say that the product $\Pi(1+a_n)$ is **divergent**. When $P_n \to 0$ we say that $\Pi(1+a_n)$ **diverges to zero.** The phrase " diverges to zero " as applied to an infinite product may at first sight seem curious, but it appears quite natural when we observe that the behaviour of the product $\Pi(1+a_n)$, $a_n > -1$, is completely determined by the behaviour of the series $\Sigma \log (1+a_n)$. This follows since

$$\log P_n = \log \{\prod_{r=1}^{n}(1+a_r)\} = \sum_{r=1}^{n} \log (1+a_r).$$

Thus, to say that the product $\Pi(1+a_n)$ diverges to zero is the same as saying that the series $\Sigma \log (1+a_n)$ diverges to $-\infty$. It should be noted that, if each of a finite number of factors has the value zero, the product is convergent if it converges when these factors are removed. In such cases the product has the *value* zero.

If the product $\prod_{n=1}^{\infty}(1+a_n)$ is convergent, P_n and P_{n-1} tend to the same limit as $n\to\infty$; that is,

$$P_n/P_{n-1} = 1+a_n \to 1,$$

whence $a_n\to 0$. Hence a necessary condition for $\Pi(1+a_n)$ to be convergent is that $a_n\to 0$.

It follows at once that, if the product $\Pi(1+a_n)$ has an infinity of negative factors, it cannot be convergent. Such products, therefore, need not concern us further. If $\Pi(1+a_n)$ has a finite number of negative factors there is an integer k such that, for $n\geqslant k+1$, $1+a_n\geqslant 0$, and we may write

$$\prod_{n=1}^{\infty}(1+a_n) = \prod_{n=1}^{k}(1+a_n) \prod_{n=k+1}^{\infty}(1+a_n).$$

It is clear, therefore, that, as regards convergence and divergence, the product $\prod_{n=1}^{\infty}(1+a_n)$ behaves in exactly the same way as the product $\prod_{n=k+1}^{\infty}(1+a_n)$. We shall thus assume throughout this chapter that $a_n>-1$ for all values of n.

50. Some Theorems on Special Types of Products. We first prove two theorems for products in which the a_n are all of the same sign.

THEOREM 51. *If $a_n\geqslant 0$ the series Σa_n and the product $\Pi(1+a_n)$ converge or diverge together.*

When $x\geqslant 0$ we have $1+x\leqslant e^x$. Thus

$$a_1+a_2+a_3\ldots+a_n<(1+a_1)(1+a_2)\ldots(1+a_n)\leqslant e^{a_1+a_2+\ldots+a_n},$$

that is,

$$A_n<P_n\leqslant e^{A_n}.$$

Since P_n and A_n are monotonic increasing functions of n the result follows.

THEOREM 52. *If $-1<a_n\leqslant 0$ the series Σa_n and the product $\Pi(1+a_n)$ converge or diverge together.*

For convenience, write $b_n = -a_n$ so that $0 \leqslant b_n < 1$. Since $1 - x \leqslant e^{-x}$ for $0 \leqslant x < 1$, we have

$$0 < P_n \leqslant e^{-(b_1 + b_2 + \ldots + b_n)}.$$

Thus if Σa_n is divergent we must have $P_n \to 0$; that is, the product diverges to zero.

Suppose, now, that Σa_n is convergent. Then, given ϵ, we can find $N = N(\epsilon)$, such that

$$0 \leqslant \sum_{\nu=N}^{\infty} b_\nu < \epsilon.$$

Also,

$$(1-b_N)(1-b_{N+1}) \geqslant 1 - b_N - b_{N+1},$$

$$(1-b_N)(1-b_{N+1})(1-b_{N+2}) \geqslant (1-b_N-b_{N+1})(1-b_{N+2})$$
$$\geqslant 1 - b_N - b_{N+1} - b_{N+2}$$

and therefore, for $n > N$,

$$(1-b_N)(1-b_{N+1}) \ldots (1-b_n) \geqslant 1 - b_N - b_{N+1} \ldots - b_n > 1 - \epsilon.$$

Clearly, P_n/P_{N-1} is monotonic decreasing and we have shown that it has a positive lower bound. It follows that P_n tends to a finite non-zero limit; that is, $\Pi(1+a_n)$ is convergent.

The following theorem provides us with an easily applied test for the convergence of an infinite product in which the a_n may be of either sign.

THEOREM 53. *If the series Σa_n^2 is convergent, then the product $\Pi(1+a_n)$ and the series Σa_n converge or diverge together.*

Since Σa_n^2 is convergent we can find N such that $|a_n| < \frac{1}{2}$ for $n > N$. For such values of n

$$|\log(1+a_n) - a_n| = \left| \frac{a_n^2}{2} - \frac{a_n^3}{3} + \ldots \right|$$
$$\leqslant \tfrac{1}{2} a_n^2 \{1 + |a_n| + |a_n^2| + \ldots\}$$
$$= \frac{a_n^2}{2(1 - |a_n|)}$$
$$< a_n^2.$$

It follows that the series $\Sigma |\log(1+a_n) - a_n|$ is convergent and therefore that the series $\Sigma\{\log(1+a_n)-a_n\}$ is convergent. That is, $\log P_n - A_n$ tends to a finite limit. The theorem therefore follows.

As illustrations of Theorems 51, 52 and 53, we observe that the products

$$\prod_{n=1}^{\infty}\left(1+\frac{1}{n}\right),\ \prod_{n=2}^{\infty}\left(1-\frac{1}{n}\right),\ \prod_{n=1}^{\infty}\left\{1+\frac{(-1)^n}{n}\right\}$$

are respectively divergent, divergent and convergent.

51. The Absolute Convergence of Infinite Products. Before defining the term " absolute convergence " as applied to an infinite product we prove a theorem of independent interest.

THEOREM 54. *If the series $\Sigma|a_n|$ is convergent, then the series $\Sigma|\log(1+a_n)|$ is also convergent.*

Since $\Sigma|a_n|$ is convergent we can find N such that $|a_n| < \frac{1}{4}$ for $n \geqslant N$. When $a_n \geqslant 0$ and $n \geqslant N$ we have, as in the proof of Theorem 53,

$$|\log(1+a_n)| = \log(1+|a_n|) \leqslant \frac{|a_n|}{1-|a_n|} \leqslant \tfrac{4}{3}|a_n|,$$

while, when $a_n<0$ and $n \geqslant N$,

$$\begin{aligned}|\log(1+a_n)| &= \log\frac{1}{1+a_n} = \log\left(1-\frac{a_n}{1+a_n}\right)\\ &= \log\left\{1+\frac{|a_n|}{1-|a_n|}\right\}\\ &\leqslant \left\{\frac{|a_n|}{1-|a_n|}\right\}\Big/\left\{1-\frac{|a_n|}{1-|a_n|}\right\}\\ &= \frac{|a_n|}{1-2|a_n|}\\ &\leqslant 2|a_n|.\end{aligned}$$

Thus, for all values of $n \geqslant N$, we have

$$|\log(1+a_n)| \leqslant 2|a_n|,$$

and the result follows from the comparison test.

We deduce at once the following theorem.

THEOREM 55. *If the product $\Pi(1+|a_n|)$ is convergent so also is the product $\Pi(1+a_n)$.*

By hypothesis and Theorem 51 the series $\Sigma|a_n|$ is convergent. Hence, by Theorem 54, the series $\Sigma|\log(1+a_n)|$ is convergent. Thus the series $\Sigma \log(1+a_n)$ and therefore the product $\Pi(1+a_n)$ are convergent.

The product $\Pi(1+a_n)$ is said to be **absolutely convergent** if the product $\Pi(1+|a_n|)$ is convergent. Theorem 55 therefore merely states that every absolutely convergent product is also convergent.

There is an analogue of Theorem 29 for infinite products.

THEOREM 56. *The factors of an absolutely convergent product $\Pi(1+a_n)$ may be rearranged in any order without affecting its convergence or its sum.*

Since the product $\Pi(1+|a_n|)$ is convergent the series $\Sigma|a_n|$ is convergent by Theorem 51. It follows from Theorem 54 that the series $\Sigma \log(1+a_n)$ is absolutely convergent. The order of the terms of this series may therefore be altered without affecting its convergence or its sum. The required result follows at once.

52. The Uniform Convergence of an Infinite Product. The infinite product $\prod_{n=1}^{\infty}\{1+a_n(x)\}$ is said to be **uniformly convergent** for $a \leqslant x \leqslant b$ if

$$P_n(x) = \prod_{n=1}^{n}\{1+a_r(x)\}$$

tends uniformly to a limit $P(x)$ for $a \leqslant x \leqslant b$.

The following theorem may often be used to test for the uniform convergence of a product.

THEOREM 57. *If the series $\Sigma|a_n(x)|$ is uniformly convergent for $a \leqslant x \leqslant b$, then so also is the product $\Pi\{1+a_n(x)\}$.*

Since the series $\Sigma|a_n(x)|$ is uniformly convergent for $a \leqslant x \leqslant b$, we can find N, independent of x, such that

$|a_n(x)|<\frac{1}{2}$ whenever $n>N$. For such values of n and $a\leqslant x\leqslant b$ we have

$$\begin{aligned}|\log\{1+a_n(x)\}| &\leqslant |a_n(x)|+\tfrac{1}{2}|a_n(x)|^2+\ldots\\ &\leqslant \frac{|a_n(x)|}{1-|a_n(x)|}\\ &\leqslant 2|a_n(x)|,\end{aligned}$$

whence it follows that $\Sigma|\log\{1+a_n(x)\}|$, and therefore $\Sigma\log\{1+a_n(x)\}$ is uniformly convergent for $a\leqslant x\leqslant b$. In other words, $\log P_n(x)$ converges uniformly to a limit which we may call $\log P(x)$; that is, $P_n(x)$ converges uniformly to a limit $P(x)$.

For example, the product $\Pi(1+x^n)$ is uniformly convergent for $|x|\leqslant\rho<1$.

We now obtain the analogue of Theorem 39.

THEOREM 58. *If the product* $\prod_{n=1}^{\infty}\{1+a_n(x)\}$ *is uniformly convergent for* $a\leqslant x\leqslant b$ *and if* $\lim_{x\to x_0} a_n(x)=a_n$, *where* $a\leqslant x_0\leqslant b$, *then*

$$\lim_{x\to x_0}\prod_{n=1}^{\infty}\{1+a_n(x)\}=\prod_{n=1}^{\infty}(1+a_n).$$

The series $\sum_{n=1}^{\infty}\log\{1+a_n(x)\}$ is uniformly convergent for $a\leqslant x\leqslant b$ so that, by Theorem 39,

$$\lim_{x\to x_0}\sum_{n=1}^{\infty}\log\{1+a_n(x)\}=\sum_{n=1}^{\infty}\log(1+a_n).$$

The result at once follows.

53. The Infinite Products for sin x and cos x. We shall show that, for all values of x,

$$\sin x = x\prod_{n=1}^{\infty}\left(1-\frac{x^2}{n^2\pi^2}\right),$$

$$\cos x = \prod_{n=1}^{\infty}\left\{1-\frac{4x^2}{(2n-1)^2\pi^2}\right\}.$$

We shall obtain the infinite product for $\sin x$ and deduce from it the infinite product for $\cos x$.

It should be noted, firstly, that the result is true if x is any multiple of π, since each side of the identity is then zero. We shall therefore suppose that x is not a multiple of π.

Secondly, we observe that, if n is an odd positive integer, $\sin nx$ is a polynomial in $\sin x$ of degree n, for, if true for $1, 3, 5, \ldots, n-2$, this is also true for n, since

$$\begin{aligned}\sin nx &= 2\sin(n-2)x\cos 2x - \sin(n-4)x \\ &= 2\sin(n-2)x\{1-2\sin^2 x\} - \sin(n-4)x.\end{aligned}$$

The assertion *is* true when $n = 1, 3$, so that it is true generally by induction.

Thirdly, $\sin nx$ vanishes when x is any multiple of π/n so that, when n is odd, we may write

$$\sin nx = K_1 \sin x \prod_{r=1}^{\frac{1}{2}(n-1)} \{\sin^2 x - \sin^2(r\pi/n)\},$$

or

$$\sin nx = K_2 \sin x \prod_{r=1}^{\frac{1}{2}(n-1)} \left\{1 - \frac{\sin^2 x}{\sin^2(r\pi/n)}\right\},$$

where K_1, K_2 are independent of x but may depend on n It now follows, on writing x for nx, that

$$\sin x = K_2 \sin(x/n) \prod_{r=1}^{\frac{1}{2}(n-1)} \left\{1 - \frac{\sin^2(x/n)}{\sin^2(r\pi/n)}\right\}.$$

In other words, for all values of x under consideration,

$$K_2 = \left\{\frac{\sin x}{\sin(x/n)}\right\} \Big/ \prod_{r=1}^{\frac{1}{2}(n-1)} \left\{1 - \frac{\sin^2(x/n)}{\sin^2(r\pi/n)}\right\}.$$

In this identity the left-hand side is independent of x and therefore so must also be the right-hand side. To determine their common value let $x \to 0$. Then clearly

$$K_2 = \lim_{x\to 0} \frac{\sin x}{\sin(x/n)} = n.$$

Thus, for all values of x under consideration and for all odd positive integers n,

$$\sin x = n \sin (x/n) \prod_{r=1}^{\frac{1}{2}(n-1)} \left\{1 - \frac{\sin^2(x/n)}{\sin^2 (r\pi/n)}\right\}$$

$$= n \sin (x/n) \prod_{r=1}^{\infty} \{1+f_r(n)\}, \quad . \quad . \quad . \quad . \quad (1)$$

where

$$f_r(n) = 0, \ \{r>\tfrac{1}{2}(n-1)\},$$

$$f_r(n) = \frac{-\sin^2 (x/n)}{\sin^2 (r\pi/n)}, \ \{r\leqslant\tfrac{1}{2}(n-1)\}.$$

From the inequality $\theta \geqslant \sin \theta \geqslant 2\theta/\pi$, $(0\leqslant\theta\leqslant\frac{1}{2}\pi)$, we have, for $n>2|x|/\pi$,

$$|f_r(n)| \leqslant \frac{x^2}{n^2} \frac{n^2\pi^2}{4r^2\pi^2} = \frac{x^2}{4r^2}$$

and the series $\Sigma \dfrac{x^2}{4r^2}$ is convergent. Thus the product on the right of (1) is uniformly convergent for all values of n and it follows from Theorem 58, on making n tend to infinity through odd integral values, that

$$\sin x = \lim_{n\to\infty} \{n \sin (x/n)\} \prod_{r=1}^{\infty} \{1+\lim_{n\to\infty} f_r(n)\}$$

$$= x \prod_{r=1}^{\infty} \left(1-\frac{x^2}{r^2\pi^2}\right).$$

To obtain the infinite product for $\cos x$ we observe that

$$\cos x = \frac{\sin 2x}{2 \sin x}$$

$$= \frac{2x \lim_{n\to\infty} \prod_{r=1}^{2n} \left(1-\frac{4x^2}{r^2\pi^2}\right)}{2x \lim_{n\to\infty} \prod_{r=1}^{n} \left(1-\frac{x^2}{r^2\pi^2}\right)}$$

$$= \lim_{n\to\infty} \left(1-\frac{4x^2}{\pi^2}\right)\left(1-\frac{4x^2}{3^2\pi^2}\right) \cdots \left\{1-\frac{4x^2}{(2n-1)^2\pi^2}\right\}$$

$$= \prod_{r=1}^{\infty} \left\{1-\frac{4x^2}{(2r-1)^2\pi^2}\right\}.$$

From the expression for sin x as an infinite product we can easily deduce the following expansion for cot x, valid for all values of x except zero or a multiple of π:—

$$\cot x = \frac{1}{x} - \sum_{r=1}^{\infty}\left(\frac{1}{r\pi - x} - \frac{1}{r\pi + x}\right).$$

We have

$$\log \sin x = \log x + \sum_{r=1}^{\infty} \log\left(1 - \frac{x^2}{r^2\pi^2}\right). \quad . \quad . \quad . \quad (2)$$

whence

$$\cot x = \frac{1}{x} - \sum_{r=1}^{\infty} \frac{2x}{r^2\pi^2 - x^2} \quad . \quad . \quad . \quad (3)$$

for all values of x for which term by term differentiation can be justified. If x lies in the interval

$$k\pi + \epsilon_1 \leqslant x \leqslant (k+1)\pi - \epsilon_2,$$

where k is zero or a positive integer, we have

$$\sum_{r=k+2}^{\infty}\left|\frac{-2x}{r^2\pi^2 - x^2}\right| \leqslant 2(k+1)\pi \sum_{r=k+2}^{\infty} \frac{1}{r^2\pi^2 - (k+1)^2\pi^2}$$
$$= \frac{2(k+1)}{\pi} \sum_{r=k+2}^{\infty} \frac{1}{r^2 - (k+1)^2},$$

and this series is convergent. We obtain a similar result when k is a negative integer. Series (3) is therefore uniformly convergent for any range of values of x which does not include a multiple of π. It follows that term by term differentiation of (2) is permissible for such values of x and that (3) holds for all values of x which are not multiples of π.

The stated result at once follows.

Example. Prove that $\sum_{n=1}^{\infty} \frac{1}{n^2} = \frac{\pi^2}{6}$.

From the series

$$\sin x = x - \frac{x^3}{3!} + \ldots,$$

we have, as $x \to 0$,

$$\frac{\sin x}{x} = 1 - \frac{x^2}{6} + O(x^4),$$

$$\log \frac{\sin x}{x} = \log \left\{1 - \frac{x^2}{6} + O(x^4)\right\}$$

$$\sim -\frac{x^2}{6}.$$

On the other hand,

$$\frac{1}{x^2} \log \frac{\sin x}{x} = \frac{1}{x^2} \sum_{n=1}^{\infty} \log\left(1 - \frac{x^2}{n^2\pi^2}\right) = -\sum_{n=1}^{\infty} \frac{1}{n^2\pi^2} + \sum_{n=1}^{\infty} O\left(\frac{x^2}{n^4}\right).$$

Let $x \to 0$. Then, from Theorem 39,

$$-\frac{1}{6} = -\sum_{n=1}^{\infty} \frac{1}{n^2\pi^2}$$

which leads to the required result.

54. The Gamma Function.* Suppose that x is neither zero nor a negative integer and that

$$P_n(x) = \frac{n^x n!}{x(x+1)\dots(x+n)} \quad . \quad . \quad . \quad (1)$$

We may write

$$P_n(x) = 1\Big/\left\{x\left(1+\frac{x}{1}\right)\left(1+\frac{x}{2}\right)\dots\left(1+\frac{x}{n}\right)e^{-x\log n}\right\}$$

$$= \frac{1}{xe^{x(1+\frac{1}{2}+\frac{1}{3}\dots+\frac{1}{n}-\log n)} \prod_{r=1}^{n}\left\{\left(1+\frac{x}{r}\right)e^{-x/r}\right\}}.$$

The series $\Sigma\left\{\log\left(1+\frac{x}{r}\right)-\frac{x}{r}\right\}$ behaves like $\Sigma \frac{x^2}{r^2}$ and is therefore convergent. Thus the infinite product $\Pi\left(1+\frac{x}{r}\right)e^{-x/r}$ is convergent and we have

$$\lim_{n\to\infty} P_n(x) = \frac{1}{xe^{\gamma x}\prod_{r=1}^{\infty}\left(1+\frac{x}{r}\right)e^{-x/r}}, \quad . \quad . \quad . \quad (2)$$

---

* See *G.*, § 37.

this limit having been shown to exist. This limit defines for all values of x except zero or a negative integer the Gamma function $\Gamma(x)$.

We proceed to obtain some properties of $\Gamma(x)$.

(i) *If x is neither zero nor a negative integer*

$$\Gamma(x+1) = x\Gamma(x).$$

We have

$$\begin{aligned}\Gamma(x+1) &= \lim_{n\to\infty} \frac{n^{x+1}n!}{(x+1)(x+2)\dots(x+n+1)} \\ &= x \lim_{n\to\infty} \frac{n}{x+n+1} \lim_{n\to\infty} \frac{n^x n!}{x(x+1)\dots(x+n)} \\ &= x\Gamma(x).\end{aligned}$$

In the particular case when x is the positive integer n we obtain, by repeated application,

$$\begin{aligned}\Gamma(n+1) &= n\Gamma(n) \\ &= n(n-1)\Gamma(n-1) \\ &\quad \cdots\cdots \\ &= n(n-1)\dots3.2.1.\Gamma(1) \\ &= n! \lim_{n\to\infty} \frac{n.n!}{1.2\dots(n+1)} \\ &= n!.\end{aligned}$$

This shows that $\Gamma(n+1)$ may be taken as a suitable definition of the symbol $n!$ where n is any real number except a negative integer.

(ii) *If x is neither zero nor an integer*

$$\Gamma(x)\Gamma(1-x) = \frac{\pi}{\sin \pi x}.$$

We have

$$\Gamma(x)\Gamma(1-x)$$

$$= \lim_{n\to\infty} \frac{n^x\, n!}{x(x+1)\ldots(x+n)} \cdot \lim_{n\to\infty} \frac{n^{1-x}\, n!}{(1-x)(2-x)\ldots(n+1-x)}$$

$$= \lim_{n\to\infty} \frac{n}{n+1-x} \cdot \lim_{n\to\infty} \frac{1}{x \prod_{r=1}^{n} \left(1-\frac{x^2}{r^2}\right)}$$

$$= \frac{1}{x \prod_{r=1}^{\infty} \left(1-\frac{x^2}{r^2}\right)}$$

$$= \frac{\pi}{\sin \pi x}.$$

When $x = \frac{1}{2}$ this result becomes

$$\{\Gamma(\tfrac{1}{2})\}^2 = \frac{\pi}{\sin \frac{1}{2}\pi} = \pi.$$

Now (2) shows that, when x is positive, $\Gamma(x)$ is also positive. It therefore follows that $\Gamma(\frac{1}{2}) = \sqrt{\pi}$.

(iii) (*Duplication Formula.*) *For all values of x for which the Gamma functions are defined*

$$\Gamma(2x) = \frac{2^{2x-1}}{\sqrt{\pi}}\, \Gamma(x)\Gamma(x+\tfrac{1}{2}).$$

We have

$$\frac{\Gamma(x)\Gamma(x+\frac{1}{2})}{\Gamma(2x)}$$

$$= \lim_{n\to\infty} \frac{n^x\, n!}{x(x+1)\ldots(x+n)} \lim_{n\to\infty} \frac{n^{x+\frac{1}{2}}\, n!}{(x+\frac{1}{2})(x+\frac{3}{2})\ldots(x+n+\frac{1}{2})}$$

$$\times \lim_{n\to\infty} \frac{2x(2x+1)\ldots(2x+2n)}{(2n)^{2x}(2n)!}$$

$$= \lim_{n\to\infty} \frac{n^{\frac{1}{2}} 2^{2n-2x+1}}{x+n+\frac{1}{2}} \frac{(n!)^2}{(2n)!}$$

$$= \lim_{n\to\infty} \frac{n^{\frac{1}{2}} 2^{2n-2x+1}}{x+n+\frac{1}{2}} \frac{\{\sqrt{(2\pi)} n^{n+\frac{1}{2}} e^{-n}\}^2}{\{\sqrt{(2\pi)}(2n)^{2n+\frac{1}{2}} e^{-2n}\}}, \text{ (Art. 33)}$$

$$= \sqrt{(2\pi)} 2^{-2x+\frac{1}{2}} \lim_{n\to\infty} \frac{n}{x+n+\frac{1}{2}}$$

$$= \frac{\sqrt{\pi}}{2^{2x-1}}.$$

Example. Prove that, if a, b, $a+b$ are not negative integers,

$$\prod_{r=1}^{\infty} \frac{r(r+a+b)}{(r+a)(r+b)} = \frac{\Gamma(1+a)\Gamma(1+b)}{\Gamma(1+a+b)}.$$

We have

$$\prod_{r=1}^{n+1} \frac{r(r+a+b)}{(r+a)(r+b)}$$

$$= \frac{(1+a+b)(2+a+b)\ldots(1+a+b+n)}{n^{1+a+b}\, n!} \cdot \frac{n^{1+a}\, n!}{(1+a)(2+a)\ldots(1+a+n)}$$

$$\times \frac{n^{1+b}\, n!}{(1+b)(2+b)\ldots(1+b+n)} \cdot \frac{n+1}{n},$$

and the result follows on making n tend to infinity.

Examples

1. Prove that

(i) $\prod_{n=1}^{\infty} 2^{n/2^n} = 4$, (ii) $\prod_{n=0}^{\infty} \{1+(\frac{1}{2})^{2^n}\} = 2$, (iii) $\prod_{n=2}^{\infty} \frac{n^3-1}{n^3+1} = \frac{2}{3}$.

2. If a_n is positive, monotonic decreasing with limit zero, show that $\Pi\{(1+a_n)^{(-1)^n}\}$ is convergent. Do the same restrictions on a_n imply the convergence of the product $\Pi\{1+(-1)^n a_n\}$?

3. Prove that the product

$$\left(1+\frac{1}{\sqrt{2}}\right)\left(1-\frac{1}{\sqrt{3}}\right)\left(1+\frac{1}{\sqrt{4}}\right)\left(1-\frac{1}{\sqrt{5}}\right)\ldots$$

diverges to zero.

4. Discuss the convergence of the products

(i) $\prod_{n=1}^{\infty}\left(1-n\sin\frac{\theta}{n^2}\right)$, (ii) $\prod_{n=1}^{\infty}\left\{1+\left(\frac{nx}{n+1}\right)^n\right\}$,

(iii) $\prod_{n=2}^{\infty}\left\{1+\frac{(-1)^n}{n^a}\right\}$, (iv) $\prod_{n=1}^{\infty}\frac{\sin\theta+n}{\cos\theta+n}$, $(-\frac{1}{2}\pi<\theta<\frac{1}{2}\pi)$,

(v) $\prod_{n=0}^{\infty}\left(\frac{x+x^{2n}}{1+x^{2n}}\right)$, (vi) $\prod_{n=1}^{\infty}\left\{\frac{1-e^{-a/n}}{\log\left(1+\frac{a}{n}\right)}\right\}$.

5. Prove that

$$\text{(i)}\ \prod_{n=1}^{\infty}\{1+2x^{2^{n-1}}\cos(2^{n-1}\theta)+x^{2^n}\} = \frac{1}{1-2x\cos\theta+x^2},\quad (-1<x<1),$$

$$\text{(ii)}\ \prod_{n=1}^{\infty}\{1+e^{-2^n\phi}\} = \tfrac{1}{2}(1+\coth\phi),\quad (\phi>0).$$

Deduce from (ii) that, when $\phi>0$,

$$\sum_{n=0}^{\infty}2^n(1-\tanh 2^n\phi) = \coth\phi-1.$$

6. Prove that

$$(1+\tfrac{1}{2})(1-\tfrac{1}{3})(1+\tfrac{1}{4})\ldots = 1,$$

but that when the factors are rearranged in the form

$$(1+\tfrac{1}{2})(1+\tfrac{1}{4})(1+\tfrac{1}{6})(1-\tfrac{1}{3})(1+\tfrac{1}{8})(1+\tfrac{1}{10})\ldots,$$

where three terms greater than unity are followed by one term less than unity, the product is equal to $\sqrt{3}$.

7. Prove that

$$\text{(i)}\ \frac{\sin\pi x}{\pi x(x+1)} = \prod_{n=1}^{\infty}\left\{\left(1-\frac{x}{n}\right)\left(1+\frac{x}{n+1}\right)\right\},$$

$$\text{(ii)}\ \pi^2\operatorname{cosec}^2\pi x = \sum_{n=-\infty}^{\infty}(x+n)^{-2}.$$

8. Prove that, for all values of x,

$$\lim_{n\to\infty}\prod_{r=n+1}^{2n}\left(1-\frac{x}{r}\right) = 2^{-x}.$$

9. Prove that, if y, $y+x$, $y-x$ are neither zero nor negative integers,

$$\prod_{n=0}^{\infty}\left\{1-\frac{x^2}{(n+y)^2}\right\}=\frac{\{\Gamma(y)\}^2}{\Gamma(y-x)\Gamma(y+x)}.$$

10. Prove that

$$\text{(i)}\ \prod_{n=1}^{\infty}\frac{4n(4n-2)}{(4n+1)(4n-3)}=\frac{\{\Gamma(\frac{1}{4})\}^2}{4\sqrt{\pi}},$$

$$\text{(ii)}\ \sum_{n=1}^{\infty}\left(\frac{1}{n}-2\log\frac{2n+2}{2n+1}\right)=\gamma-\log\tfrac{1}{4}\pi.$$

11. Prove that

$$\cos\frac{\pi x}{3}+\frac{1}{\sqrt{3}}\sin\frac{\pi x}{3}=\left(1+\frac{x}{1}\right)\left(1-\frac{x}{2}\right)\left(1+\frac{x}{4}\right)\left(1-\frac{x}{5}\right)$$
$$\times\left(1+\frac{x}{7}\right)\left(1-\frac{x}{8}\right)\ldots$$

12. Show that, with certain restrictions on the values of x,

$$\text{(i)}\ \frac{d}{dx}\log\Gamma(x+1)=-\gamma+\sum_{r=1}^{\infty}\left(\frac{1}{r}-\frac{1}{x+r}\right),$$

$$\text{(ii)}\ \Gamma(x+\tfrac{1}{4})\Gamma(x+\tfrac{1}{2})\Gamma(x+\tfrac{3}{4})\Gamma(x+1)=(2\pi)^{3/2}2^{-8x-1}\Gamma(4x+1).$$

13. Prove that

$$\prod_{r=1}^{n-1}\Gamma\left(\frac{r}{n}\right)=\frac{(2\pi)^{\frac{1}{2}(n-1)}}{\sqrt{n}}.$$

ANSWERS. 2. No; see Example 3. 4. (i) Diverges to zero if $\theta>0$, converges if $\theta=0$, diverges if $\theta<0$; (ii) converges for $|x|<1$; (iii) converges for $a>\frac{1}{2}$; (iv) converges for $\theta=\frac{1}{4}\pi$; (v) converges for $|x|>1$ and for $x=1$, diverges to zero for $-1\leqslant x<1$; (vi) convergent, [a must be >-1]. **12.** x must not have any value which makes the argument of one of the Gamma functions zero or a negative integer.

CHAPTER IX

DOUBLE SERIES

55. Introduction. Suppose that we have the array of numbers

$$\begin{array}{ccccc} a_{11} & a_{12} & a_{13} & a_{14} & \cdots \\ a_{21} & a_{22} & a_{23} & a_{24} & \cdots \\ a_{31} & a_{32} & a_{33} & a_{34} & \cdots \\ a_{41} & a_{42} & a_{43} & a_{44} & \cdots \\ & & \cdots\cdot & & \end{array}$$

We wish to consider the series whose terms are the members of this array. The series is called a **double series** and is denoted by $\sum\limits_{m,n=1}^{\infty} a_{mn}$. In defining what we mean by the sum of such a series we are at once confronted by a difficulty. For example, the following four definitions of the sum might be regarded as quite reasonable.

$$\lim_{N\to\infty} \sum_{\nu=2}^{N} \Big(\sum_{m+n=\nu} a_{mn}\Big), \qquad (1)$$

$$\lim_{N\to\infty} \Big(\sum_{m=N,\, n<N} a_{mn} + \sum_{m<N,\, n=N} a_{mn} + a_{NN}\Big), \qquad (2)$$

$$\lim_{M\to\infty} \Big\{ \sum_{m=1}^{M} \Big(\lim_{N\to\infty} \sum_{n=1}^{N} a_{mn}\Big)\Big\}, \qquad (3)$$

$$\lim_{N\to\infty} \Big\{ \sum_{n=1}^{N} \Big(\lim_{M\to\infty} \sum_{m=1}^{M} a_{mn}\Big)\Big\} \qquad (4)$$

In the first* we are summing by "triangles," in the second†

* $\sum\limits_{m+n=\nu} a_{m,n} = a_{1,\nu-1} + a_{2,\nu-2} + \ldots + a_{\nu-1,1}$.

† $\sum\limits_{m=N,\, n<N} a_{m,n} = a_{N,1} + a_{N,2} + \ldots + a_{N,N-1}$.

by " squares," in the third by " rows " and in the last by " columns." Clearly these methods of defining the sum of the double series are only four of an infinite number which could be devised.

Naturally, we wish our definition of the sum of a double series to conform as closely as possible to the definition of the sum of a single series. This analogy may be preserved by starting at the top left-hand corner of the array and taking successive groups of terms, where each group consists only of a finite number of terms of the series and contains all the elements of the preceding group, and then examining the limit of the p-th group as p tends to infinity. These successive groups correspond in fact to successive partial sums in the case of single series. Generally speaking, the limit of the p-th group will depend on the system by means of which the groups are formed. When, however, the limit is finite and independent of the system of grouping we say that the double series is **convergent** and that the limit in question is the sum of the series. In all other cases the series is said to be **divergent**. We use the term **properly divergent** in the case of a series where the p-th group tends to $+\infty$ or to $-\infty$ for all possible systems of grouping.

It will be noted that the third and fourth definitions above are not included in this general definition since our groups in these two cases do not consist of a finite number of terms. If (3) is finite we say that the **repeated series** $\sum_{m=1}^{\infty} \sum_{n=1}^{\infty} a_{mn}$ is convergent to the value of (3), and if (4) is finite we say that the repeated series $\sum_{n=1}^{\infty} \sum_{m=1}^{\infty} a_{mn}$ is convergent to the value of (4).

The following preliminary theorem will serve to illustrate these definitions.

THEOREM 59.

(i) *If* $\sum_{m,n=1}^{\infty} a_{mn}$ *converges to the sum* α *and if* $\sum_{m,n=1}^{\infty} b_{mn}$ *converges to the sum* β, *then* $\sum_{m,n=1}^{\infty} (a_{mn}+b_{mn})$ *converges to the sum* $(\alpha+\beta)$.

(ii) *If* $\sum_{m,n=1}^{\infty} a_{mn}$ *converges to the sum* α *and if* c *is independent of* m *and* n, *then* $\sum_{m,n=1}^{\infty} ca_{mn}$ *converges to the sum* $c\alpha$.

Similar results are true for repeated series.

Write $c_{mn} = a_{mn}+b_{mn}$. Take any system of grouping $G_1, G_2, \ldots, G_p, \ldots$ and let g_p, g_p', g_p'' denote the sums of all the terms in the group G_p for the series Σa_{mn}, Σb_{mn} and Σc_{mn} respectively. Clearly $g_p'' = g_p+g_p'$. But g_p and g_p' tend respectively to α and β as p tends to infinity. Hence $g_p''\to\alpha+\beta$, and this holds no matter what system of grouping is adopted. Result (i) therefore follows.

We leave to the reader the proof of (ii) and the consideration of the case of repeated series.

56. Double Series whose Terms are Non-negative.

THEOREM 60. *If, for all values of* m *and* n, $a_{mn}\geqslant 0$, *then the double series* $\sum_{m,n=1}^{\infty} a_{mn}$ *and the repeated series* $\sum_{m=1}^{\infty}\sum_{n=1}^{\infty} a_{mn}$, $\sum_{n=1}^{\infty}\sum_{m=1}^{\infty} a_{mn}$ *either all converge to a finite sum* α *or else they are all properly divergent.*

We prove first that the double series either converges to a sum α or is properly divergent.

Consider any method of grouping the terms of the series. Let the successive groups be denoted by $G_1, G_2, \ldots, G_p, \ldots$ and let the sums of the terms in these groups be denoted by $g_1, g_2, \ldots, g_p, \ldots$ respectively. Then, in accordance with the definition of our system of grouping, and since $a_{mn} \geqslant 0$, we have

$$g_1 \leqslant g_2 \leqslant g_3 \leqslant \ldots .$$

Suppose that the sums of all selections of terms from the double series, finite in number, are bounded and have upper bound a. Then clearly, for all values of p, $g_p \leqslant a$. On the other hand, given ϵ, there is one finite sum at least which is greater than $a-\epsilon$. By choosing a large enough value of p, say p_1, we can include all the terms of this finite sum in the group G_{p_1}. Thus $g_{p_1} > a-\epsilon$ and *a fortiori* $g_p > a-\epsilon$ whenever $p \geqslant p_1$. Hence, as $p \to \infty$, $g_p \to a$, and, since this is independent of the system of grouping, it follows that in this case the double series converges to the sum a.

Suppose now that there is no upper bound for all the finite sums of the terms of the series. Then, given any positive number K, there is at least one finite sum which is greater than K. As before, we can find a value p_1 of p such that G_{p_1} contains all the terms of this finite sum. Hence $g_p > K$ for $p \geqslant p_1$ and the double series therefore diverges to $+\infty$.

We have now to consider the case of the two repeated series. It will clearly be sufficient to prove that the two series

$$\sum_{p=1}^{\infty} \left(\sum_{m+n=p} a_{mn} \right), \quad \sum_{m=1}^{\infty} \sum_{n=1}^{\infty} a_{mn}$$

converge or diverge together and that, when convergent, their sums are equal.

Suppose first that the double series converges to the sum a. Let $c_p = \sum_{m+n=p} a_{mn}$. Then, clearly, for any fixed value of m

$$a_{m1} + a_{m2} + \ldots \quad \leqslant c_{m+1} + c_{m+2} + \ldots .$$

Since the double series is convergent the series on the

right is convergent and it therefore follows that, for each fixed value of m, the series $\sum_{n=1}^{\infty} a_{mn}$ is convergent.

Write
$$C_\mu = \sum_{p=2}^{\mu} c_p,\quad C'_\mu = \sum_{m=1}^{\mu}\left(\sum_{n=1}^{\infty} a_{mn}\right).$$

Then it is clear that $C_\mu \leqslant C'_\mu$, whence $a \leqslant \lim_{\mu\to\infty} C'_\mu$. Again, we may write

$$C'_\mu = \sum_{m=1}^{\mu}\left\{\sum_{n=1}^{\nu} a_{mn} + r_{m,\nu+1}\right\},$$

where $r_{m,\nu+1} = \sum_{n=\nu+1}^{\infty} a_{mn}$. Given ϵ, we can determine ν_k such that, for $\nu > \nu_k$,

$$|r_{k,\nu+1}| < \epsilon/\mu,\ (k = 1, 2, \dots \mu).$$

Let ν be fixed and greater than $\operatorname*{Max}_{k=1,2,\dots\mu} \nu_k$. Then

$$\begin{aligned} C'_\mu &\leqslant \sum_{m=1}^{\mu}\sum_{n=1}^{\nu} a_{mn} + \epsilon \\ &\leqslant C_{\mu+\nu} + \epsilon. \end{aligned}$$

It follows that $\lim_{\mu\to\infty} C'_\mu \leqslant a+\epsilon$. But ϵ is arbitrary, so that $\lim C'_\mu \leqslant a$. We have already proved that $\lim C'_\mu \geqslant a$. The repeated series therefore converges to the sum a.

Now suppose that the double series is properly divergent. Then either $\sum_{n=1}^{\infty} a_{mn}$ diverges to $+\infty$ for some value of m, or $\sum_{n=1}^{\infty} a_{mn}$ converges for every value of m. If the former is true the repeated series diverges to $+\infty$ and no further proof is required. If the latter is true we prove exactly as before that $C_\mu \leqslant C'_\mu$, whence $C'_\mu \to +\infty$.

The theorem is therefore completely proved.

We now obtain the analogue for double series of the comparison test.

THEOREM 61. *If, for all values of m and n, $a_{mn} \geqslant b_{mn} \geqslant 0$, and if the series $\sum_{m,n=1}^{\infty} a_{mn}$ is convergent, so also is the series $\sum_{m,n=1}^{\infty} b_{mn}$. A similar result also holds for repeated series.*

Take any system of grouping $G_1, G_2, \ldots$ for the series Σa_{mn} and Σb_{mn} and let g_p, g'_p be the sum of the terms of Σa_{mn} and Σb_{mn} respectively in G_p. Clearly $g'_p \leqslant g_p$. Now g_p tends to a finite limit and g'_p is monotonic increasing. It follows that g'_p tends to a finite limit. For double series the theorem is therefore proved. The result is obvious in the case of repeated series.

Example. Examine for convergence the repeated series

$$\sum_{m=1}^{\infty} \sum_{n=1}^{\infty} \frac{1}{m^{\alpha}+n^{\alpha}}.$$

This series converges or diverges with the double series $\sum_{m,n=1}^{\infty} \frac{1}{m^{\alpha}+n^{\alpha}}$ and, in particular, with the series

$$\sum_{p=1}^{\infty} \left\{ \sum_{m+n=p} \frac{1}{m^{\alpha}+n^{\alpha}} \right\}.$$

If $\alpha<1$, $m+n=p$, we have

$$2p^{\alpha} > m^{\alpha}+n^{\alpha} \geqslant 2(\tfrac{1}{2}p)^{\alpha}$$

whence

$$\frac{p-1}{2p^{\alpha}} < \sum_{m+n=p} \frac{1}{m^{\alpha}+n^{\alpha}} \leqslant \frac{p-1}{2(\frac{1}{2}p)^{\alpha}}.$$

It therefore follows that the double series converges or diverges with the series $\Sigma 1/p^{\alpha-1}$. Hence the double series, and therefore the repeated series, is convergent if $\alpha>2$ and properly divergent if $\alpha \leqslant 2$.

57. The Absolute Convergence of a Double Series. The definition of the absolute convergence of a double series is analogous to the corresponding definition for single series. We first prove the following theorem.

THEOREM 62. *If the series* $\sum_{m,n=1}^{\infty} |a_{mn}|$ *is convergent, then so is the series* $\sum_{m,n=1}^{\infty} a_{mn}$. *A similar result holds for repeated series.*

Let

$$b_{mn} = a_{mn},\ (a_{mn} \geqslant 0);\quad c_{mn} = -a_{mn},\ (a_{mn} \leqslant 0).$$
$$\quad\ = 0,\ (a_{mn} < 0) \qquad\qquad = 0,\ (a_{mn} > 0)$$

Then

$$a_{mn} = b_{mn} - c_{mn},\ |a_{mn}| = b_{mn} + c_{mn}.$$

Each of the series $\sum_{m,n=1}^{\infty} b_{mn}$, $\sum_{m,n=1}^{\infty} c_{mn}$ is a convergent double series of non-negative terms, by comparison with the series $\sum_{m,n=1}^{\infty} |a_{mn}|$. It follows from Theorem 59 that $\sum_{m,n=1}^{\infty} (b_{mn} - c_{mn}) = \sum_{m,n=1}^{\infty} a_{mn}$ is convergent.

In the case of repeated series the proof is similar.

If the double series $\sum_{m,n=1}^{\infty} |a_{mn}|$ is convergent, then we say that the series $\sum_{m,n=1}^{\infty} a_{mn}$ is absolutely convergent. If the repeated series $\sum_{m=1}^{\infty} \sum_{n=1}^{\infty} |a_{mn}|$, $\sum_{n=1}^{\infty} \sum_{m=1}^{\infty} |a_{mn}|$ are convergent, then we say that the repeated series $\sum_{m=1}^{\infty} \sum_{n=1}^{\infty} a_{mn}$, $\sum_{n=1}^{\infty} \sum_{m=1}^{\infty} a_{mn}$ are absolutely convergent.

As in the case of single series most properties of double series of non-negative terms remain true for series whose terms are not all of the same sign but which are absolutely convergent. In particular, if one of the series $\sum_{m=1}^{\infty} \sum_{n=1}^{\infty} a_{mn}$, $\sum_{n=1}^{\infty} \sum_{m=1}^{\infty} a_{mn}$, $\sum_{m,n=1}^{\infty} a_{mn}$ is absolutely convergent, then so are the other two and the sums of all three series are the same. We leave this general result to the consideration of the reader, although it will in part be proved in the next article.

58. The Interchange of the Order of Summation for Repeated Series. We now consider in a little more detail an important special problem relating to repeated series. We wish to investigate under what conditions we are entitled to change the order of summation in the series $\sum_{m=1}^{\infty}\sum_{n=1}^{\infty} a_{mn}$. We have already discussed this question in certain particular cases. For example, we have proved that we are entitled to change the order when $a_{mn} \geqslant 0$ and have stated that we can also do so when either of the repeated series is absolutely convergent. We shall now prove the latter result.

THEOREM 63. *If either of the series* $\sum_{m=1}^{\infty}\sum_{n=1}^{\infty} a_{mn}$, $\sum_{n=1}^{\infty}\sum_{m=1}^{\infty} a_{mn}$ *is absolutely convergent, then so is the other and their sums are the same.*

Suppose that the first series is absolutely convergent. This is the same as saying that, for each value of m, the series $\sum_{n=1}^{\infty} |a_{mn}|$ converges to a sum σ_m and that the series $\sum_{m=1}^{\infty} \sigma_m$ is convergent. The absolute convergence of the second series follows at once from Theorem 60. We therefore confine ourselves to proving that the sums of the two series are the same.

We may write

$$\begin{aligned}\sum_{m=1}^{\infty}\sum_{n=1}^{\infty} a_{mn} &= \sum_{m=1}^{\infty}\sum_{n=1}^{N} a_{mn} + \sum_{m=1}^{\infty}\sum_{n=N+1}^{\infty} a_{mn} \\ &= \sum_{n=1}^{N}\sum_{m=1}^{\infty} a_{mn} + \sum_{m=1}^{\infty}\sum_{n=N+1}^{\infty} a_{mn}\end{aligned}$$

since N is finite and $\sum_{m=1}^{\infty} a_{mn}$ converges for each value of n.

The theorem will then be proved if we show that

$$\lim_{N\to\infty} \sum_{m=1}^{\infty} \sum_{n=N+1}^{\infty} a_{mn} = 0.$$

Let $\rho_m(N) = \sum_{n=N+1}^{\infty} a_{mn}$. Then

$$|\rho_m(N)| \leqslant \sum_{n=N+1}^{\infty} |a_{mn}| \leqslant \sigma_m$$

and therefore, by Theorem 36, the series $\sum_{m=1}^{\infty} \rho_m(N)$ is uniformly convergent for all values of N. Hence, by Theorem 39,

$$\begin{aligned} \lim_{N\to\infty} \sum_{m=1}^{\infty} \sum_{n=N+1}^{\infty} a_{mn} &= \lim_{N\to\infty} \sum_{m=1}^{\infty} \rho_m(N) \\ &= \sum_{m=1}^{\infty} \{\lim_{N\to\infty} \rho_m(N)\} \\ &= 0, \end{aligned}$$

since the series $\sum_{n=1}^{\infty} a_{mn}$ converges for each value of m.

A slightly more general theorem of the same type is the following.

THEOREM 64. *If*

$$\rho_m(N) = \sum_{n=N+1}^{\infty} a_{mn}$$

and if, for all values of N, $|\rho_m(N)| < \sigma_m$ where the series $\sum_{m=1}^{\infty} \sigma_m$ is convergent then the convergence of the series $\sum_{m=1}^{\infty} \sum_{n=1}^{\infty} a_{mn}$ implies the convergence of the series $\sum_{n=1}^{\infty} \sum_{m=1}^{\infty} a_{mn}$ and the sums of the two series are the same.

We observe that, for each fixed value of n,

$$|a_{mn}| = |\rho_m(n-1) - \rho_m(n)| \leqslant 2\sigma_m$$

so that the series $\sum_{m=1}^{\infty} a_{mn}$ is convergent for all values of n.

Repetition of the proof of Theorem 63 now yields the desired result.

Example. Prove that

$$\sum_{m=1}^{\infty}\sum_{n=1}^{\infty}\frac{(-1)^n}{(m+n^2)(m+n^2-1)}=-\frac{\pi^2}{12}.$$

Denote the given series by S, and by S' the series

$$\sum_{n=1}^{\infty}(-1)^n\sum_{m=1}^{\infty}\frac{1}{(n^2+m)(n^2+m-1)}.$$

Now

$$\frac{1}{(n^2+m)(n^2+m-1)}=\frac{1}{n^2+m-1}-\frac{1}{n^2+m}$$

so that the series $\sum_{m=1}^{\infty}\frac{1}{(n^2+m)(n^2+m-1)}$ converges for each value of n to the sum $1/n^2$. The series S' is therefore absolutely convergent to the sum of the series $\sum_{n=1}^{\infty}(-1)^n/n^2$. Hence, by Theorem 63, the sum of the series S is equal to $\sum_{n=1}^{\infty}(-1)^n/n^2$. Now

$$\sum_{n=1}^{\infty}(-1)^n/n^2=\sum_{n=1}^{\infty}1/n^2-\tfrac{1}{2}\sum_{n=1}^{\infty}1/n^2=-\frac{\pi^2}{12}.$$

Examples

1. Examine the convergence of the series

$$\sum_{m,n=2}^{\infty}\frac{1}{(am^\alpha+bn^\alpha)(\log mn)^\beta}.$$

2. Examine the convergence of the series $\sum_{m,n=1}^{\infty}\frac{1}{m^\alpha n^\beta}$ and prove that

$$\sum_{m,n=1}^{\infty}{}'\frac{1}{m^2n^2}=\frac{\pi^4}{120},$$

where the dash denotes that those terms for which $m=n$ are omitted from the summation.

3. Show that, if α and β are greater than 1, the series

$$\sum_{m,n=1}^{\infty} \frac{1}{m^{\alpha}+n^{\beta}}$$

converges if $\beta>\alpha/(\alpha-1)$ and diverges if $\beta\leqslant\alpha/(\alpha-1)$. What happens if α or β or both are less than 1?

[Consider the corresponding repeated series and use Theorem 16.]

4. Show that

$$\frac{1}{1+x+x^2} = \sum_{n=0}^{\infty} \frac{\sin\{2\pi(n+1)/3\}}{\sin(2\pi/3)} x^n = 4\sum_{n=0}^{\infty} \frac{(-1)^n}{3^{n+1}}(1+2x)^{2n}$$

stating the range of validity of each expansion. Deduce the sum of the series

$$\sum_{n=m}^{\infty} \frac{(-1)^n}{3^n} \cdot \frac{(2n)!}{(2n-r)!\,r!}$$

for any positive integer r, where $m = \frac{1}{2}r$ or $\frac{1}{2}(r+1)$ according as r is even or odd.

5. Prove that, if $|x|<1$,

$$\sum_{r=1}^{\infty} \frac{x^r}{(1-x^{2r})^2} = \sum_{r=1}^{\infty} \frac{rx^{2r-1}}{1-x^{2r-1}}.$$

For what values of x are the two series convergent? Show that if $|x|>1$ the first series is equal to

$$\sum_{r=1}^{\infty} \frac{r}{x^{2r+1}-1}.$$

ANSWERS. 1. Convergent if a and b have the same sign and if $\alpha>2$. 2. Convergent for $\alpha>1$, $\beta>1$. 3. The series diverges. 4. $|x|<1$ for the first expansion, $-(\sqrt{3}+1)<2x<(\sqrt{3}-1)$ for the second. 5. The first series converges for all values of x except ±1, the second for $|x|<1$.

INDEX

The numbers refer to the pages

Abel's lemma, 65
Abel's test, 61
Abel's theorem, 80
Absolute convergence, 58, 59
96, 97, 113

Binomial series, 27, 50, 81
Bounds of a function, 2

Cauchy's test, 44, 45
Circular functions, 23, 27, 98, 100
Closed interval, 2
Comparison tests, 40, 41, 42
Complex limits, 63
Complex series, 64
Conditional convergence, 60, 61
Constant, Euler's, 51
Continuity, of a function, 10
of series, 74, 80
Convergence, absolute, 58, 59, 96, 97, 113
conditional, 60, 61
general principle of, 29
of integrals, 12
of products, 94, 95, 96
of sequences, 25
of series, 26
uniform, 68, 69, 70, 71, 72, 80
Cosine, infinite product for, 98
series for, 27
Cotangent, series of fractions for, 101

d'Alembert's test, 43
Differentiation, 11
of series, 76, 80
Dirichlet's test, 61

Divergence, of integrals, 12
of products, 93
of sequences, 25
of series, 26, 109
Double series, 108, 110, 113
absolute convergence of, 113
convergent, 109
divergent, 109
whose terms are non-negative, 110
Duplication formula for the Gamma function, 103

Euler's constant, 51
Expansions, 12
Exponential function, the, 18, 19, 27, 89
Exponential limits, 21
Exponential series, the, 27

Factorials, 51, 103
Function, bounds of a, 2
exponential, 18, 19, 27, 89
Gamma, 102
hypergeometric, 50
limits of a, 3, 4, 5, 8
logarithmic, 18, 20, 26
Functions, 1
circular, 23, 27, 98, 100
continuous, 10
hyperbolic, 22, 27
monotonic, 8
series of, 69

Gamma function, the, 102
duplication formula for, 104
expressions for factorials in terms of, 103
infinite product for, 102
properties of, 103

Gauss's test, 48
General principle of convergence, 29
General series, 58
Gillespie, 12
Gregory's series, 81
Grouping of terms in infinite series, 33

Hyperbolic functions, the, 22
  series for, 27
Hypergeometric series, the, 50

Infinite products, 93
Infinite series, 26, 31, 32, 33
Integrals, convergent, 12
  divergent, 12
  oscillating, 13
Integral test for convergence, the, 38
Integration, 12
  of series, 75, 80
Interval, closed, 2
  open, 2
Inverse tangent, series for, 81

Kummer's test, 46

Lemma, Abel's, 65
Limits, complex, 63
Limits of functions, 3, 4, 5, 8, 21, 22
  upper and lower, 8, 9, 10
Logarithmic function, the, 18, 20, 26
Logarithmic series, the, 21, 26
Lower bound of a function, 2
Lower limit of a function, 8, 9, 10

Maclaurin's expansion, 12, 82
Monotonic functions, 8
M-test, Weierstrass's, 70
Multiplication of series, 86, 87, 88, 89, 90

o, *O* notation, 14
Open interval, 2
Oscillating, integral, 13
  sequence, 25
  series, 26

Partial sum of a series, 26
Power series, 79
  Abel's theorem for, 80
  continuity of, 80
  differentiation of, 80
  integration of, 80
  Maclaurin's expansion for, 82
  radius of convergence of, 79
  uniform convergence of, 80
Power, the generalised, 18
Products, infinite, 93
  absolute convergence of, 95, 96
  convergence of, 93, 94, 96
  divergence of, 93
  for sin x and cos x, 98
  for the Gamma function, 102
  uniform convergence of, 97
Proper divergence, 13, 25, 26, 109

Raabe's test, 48
Radius of convergence, 79
Ratio test, 43, 45
Real series, 58
Real variable, 1
Rearrangement of the terms of an infinite series, 36, 59, 62
Repeated series, 109, 110
  interchange of order of summation of, 115, 116
Riemann integral, the, 12
Riemann's theorem, 62

Sequences, 25
Series, binomial, 27, 50, 81
  cosine, 27
  exponential, 27
  for the hyperbolic functions, 27
  for the inverse tangent, 81
  Gregory's, 81
  hypergeometric, 50
  logarithmic, 21, 26
  sine, 27
Series, infinite, 26, 31, 32, 33
  absolute convergence of, 58, 59
  Cauchy's test for, 44, 45

Series, infinite, comparison tests for, 40, 41, 42
conditional convergence of, 60, 61
continuity of, 74, 80
convergence of, 26, 29
differentiation of, 76, 80
double, 108, 110, 113
Gauss's test for, 48
general, 58
integral test for, 38
integration of, 75, 80
Kummer's test for, 46
multiplication of, 86, 87, 88, 89, 90
of complex terms, 64
of functions, 69
of non-negative terms, 36, 86, 110, 113
power, 79
Raabe's test for, 48
radius of convergence of power, 79
ratio test for, 43, 45
real, 58
rearrangement of terms of, 36, 59, 62
repeated, 109, 110, 115, 116
uniform convergence of, 69, 73, 80
Sine, infinite product for, 98
Sine series, 27
Stirling's approximation, 51
Sum, partial, of a series, 26

Taylor's expansion, 12
Test, Abel's, 6
Cauchy's, 44, 45
Dirichlet's, 61
Gauss's, 48
Kummer's, 46
Raabe's, 48
the integral, 38
the ratio, 43, 45
Tests, comparison, 40, 41, 42
for absolute convergence, 59
for uniform convergence, 70, 71, 72

Uniform convergence, 68
of products, 97
of series, 69, 80
Uniformly convergent series, 69, 80
continuity of, 74, 80
differentiation of, 76, 80
integration of, 75, 80
properties of, 73
tests for, 70, 71, 72
Upper bound, 2
Upper limit, 8, 9, 10

Variable, real, 1

Weierstrass's *M*- test 70

Zero, divergence to, 93

A CATALOG OF SELECTED

DOVER BOOKS

IN SCIENCE AND MATHEMATICS

Mathematics

FUNCTIONAL ANALYSIS (Second Corrected Edition), George Bachman and Lawrence Narici. Excellent treatment of subject geared toward students with background in linear algebra, advanced calculus, physics and engineering. Text covers introduction to inner-product spaces, normed, metric spaces, and topological spaces; complete orthonormal sets, the Hahn-Banach Theorem and its consequences, and many other related subjects. 1966 ed. 544pp. 6⅛ x 9¼. 0-486-40251-7

ASYMPTOTIC EXPANSIONS OF INTEGRALS, Norman Bleistein & Richard A. Handelsman. Best introduction to important field with applications in a variety of scientific disciplines. New preface. Problems. Diagrams. Tables. Bibliography. Index. 448pp. 5⅜ x 8½. 0-486-65082-0

VECTOR AND TENSOR ANALYSIS WITH APPLICATIONS, A. I. Borisenko and I. E. Tarapov. Concise introduction. Worked-out problems, solutions, exercises. 257pp. 5⅜ x 8¼. 0-486-63833-2

AN INTRODUCTION TO ORDINARY DIFFERENTIAL EQUATIONS, Earl A. Coddington. A thorough and systematic first course in elementary differential equations for undergraduates in mathematics and science, with many exercises and problems (with answers). Index. 304pp. 5⅜ x 8½. 0-486-65942-9

FOURIER SERIES AND ORTHOGONAL FUNCTIONS, Harry F. Davis. An incisive text combining theory and practical example to introduce Fourier series, orthogonal functions and applications of the Fourier method to boundary-value problems. 570 exercises. Answers and notes. 416pp. 5⅜ x 8½. 0-486-65973-9

COMPUTABILITY AND UNSOLVABILITY, Martin Davis. Classic graduate-level introduction to theory of computability, usually referred to as theory of recurrent functions. New preface and appendix. 288pp. 5⅜ x 8½. 0-486-61471-9

ASYMPTOTIC METHODS IN ANALYSIS, N. G. de Bruijn. An inexpensive, comprehensive guide to asymptotic methods–the pioneering work that teaches by explaining worked examples in detail. Index. 224pp. 5⅜ x 8½ 0-486-64221-6

APPLIED COMPLEX VARIABLES, John W. Dettman. Step-by-step coverage of fundamentals of analytic function theory–plus lucid exposition of five important applications: Potential Theory; Ordinary Differential Equations; Fourier Transforms; Laplace Transforms; Asymptotic Expansions. 66 figures. Exercises at chapter ends. 512pp. 5⅜ x 8½. 0-486-64670-X

INTRODUCTION TO LINEAR ALGEBRA AND DIFFERENTIAL EQUATIONS, John W. Dettman. Excellent text covers complex numbers, determinants, orthonormal bases, Laplace transforms, much more. Exercises with solutions. Undergraduate level. 416pp. 5⅜ x 8½. 0-486-65191-6

RIEMANN'S ZETA FUNCTION, H. M. Edwards. Superb, high-level study of landmark 1859 publication entitled "On the Number of Primes Less Than a Given Magnitude" traces developments in mathematical theory that it inspired. xiv+315pp. 5⅜ x 8½. 0-486-41740-9

CALCULUS OF VARIATIONS WITH APPLICATIONS, George M. Ewing. Applications-oriented introduction to variational theory develops insight and promotes understanding of specialized books, research papers. Suitable for advanced undergraduate/graduate students as primary, supplementary text. 352pp. 5⅜ x 8½. 0-486-64856-7

COMPLEX VARIABLES, Francis J. Flanigan. Unusual approach, delaying complex algebra till harmonic functions have been analyzed from real variable viewpoint. Includes problems with answers. 364pp. 5⅜ x 8½. 0-486-61388-7

AN INTRODUCTION TO THE CALCULUS OF VARIATIONS, Charles Fox. Graduate-level text covers variations of an integral, isoperimetrical problems, least action, special relativity, approximations, more. References. 279pp. 5⅜ x 8½. 0-486-65499-0

COUNTEREXAMPLES IN ANALYSIS, Bernard R. Gelbaum and John M. H. Olmsted. These counterexamples deal mostly with the part of analysis known as "real variables." The first half covers the real number system, and the second half encompasses higher dimensions. 1962 edition. xxiv+198pp. 5⅜ x 8½. 0-486-42875-3

CATASTROPHE THEORY FOR SCIENTISTS AND ENGINEERS, Robert Gilmore. Advanced-level treatment describes mathematics of theory grounded in the work of Poincaré, R. Thom, other mathematicians. Also important applications to problems in mathematics, physics, chemistry and engineering. 1981 edition. References. 28 tables. 397 black-and-white illustrations. xvii + 666pp. 6⅛ x 9¼. 0-486-67539-4

INTRODUCTION TO DIFFERENCE EQUATIONS, Samuel Goldberg. Exceptionally clear exposition of important discipline with applications to sociology, psychology, economics. Many illustrative examples; over 250 problems. 260pp. 5⅜ x 8½. 0-486-65084-7

NUMERICAL METHODS FOR SCIENTISTS AND ENGINEERS, Richard Hamming. Classic text stresses frequency approach in coverage of algorithms, polynomial approximation, Fourier approximation, exponential approximation, other topics. Revised and enlarged 2nd edition. 721pp. 5⅜ x 8½. 0-486-65241-6

INTRODUCTION TO NUMERICAL ANALYSIS (2nd Edition), F. B. Hildebrand. Classic, fundamental treatment covers computation, approximation, interpolation, numerical differentiation and integration, other topics. 150 new problems. 669pp. 5⅜ x 8½. 0-486-65363-3

THREE PEARLS OF NUMBER THEORY, A. Y. Khinchin. Three compelling puzzles require proof of a basic law governing the world of numbers. Challenges concern van der Waerden's theorem, the Landau-Schnirelmann hypothesis and Mann's theorem, and a solution to Waring's problem. Solutions included. 64pp. 5⅜ x 8½. 0-486-40026-3

THE PHILOSOPHY OF MATHEMATICS: AN INTRODUCTORY ESSAY, Stephan Körner. Surveys the views of Plato, Aristotle, Leibniz & Kant concerning propositions and theories of applied and pure mathematics. Introduction. Two appendices. Index. 198pp. 5⅜ x 8½. 0-486-25048-2

CATALOG OF DOVER BOOKS

TENSOR CALCULUS, J.L. Synge and A. Schild. Widely used introductory text covers spaces and tensors, basic operations in Riemannian space, non-Riemannian spaces, etc. 324pp. 5⅜ x 8¼. 0-486-63612-7

ORDINARY DIFFERENTIAL EQUATIONS, Morris Tenenbaum and Harry Pollard. Exhaustive survey of ordinary differential equations for undergraduates in mathematics, engineering, science. Thorough analysis of theorems. Diagrams. Bibliography. Index. 818pp. 5⅜ x 8½. 0-486-64940-7

INTEGRAL EQUATIONS, F. G. Tricomi. Authoritative, well-written treatment of extremely useful mathematical tool with wide applications. Volterra Equations, Fredholm Equations, much more. Advanced undergraduate to graduate level. Exercises. Bibliography. 238pp. 5⅜ x 8½. 0-486-64828-1

FOURIER SERIES, Georgi P. Tolstov. Translated by Richard A. Silverman. A valuable addition to the literature on the subject, moving clearly from subject to subject and theorem to theorem. 107 problems, answers. 336pp. 5⅜ x 8½. 0-486-63317-9

INTRODUCTION TO MATHEMATICAL THINKING, Friedrich Waismann. Examinations of arithmetic, geometry, and theory of integers; rational and natural numbers; complete induction; limit and point of accumulation; remarkable curves; complex and hypercomplex numbers, more. 1959 ed. 27 figures. xii+260pp. 5⅜ x 8½. 0-486-63317-9

POPULAR LECTURES ON MATHEMATICAL LOGIC, Hao Wang. Noted logician's lucid treatment of historical developments, set theory, model theory, recursion theory and constructivism, proof theory, more. 3 appendixes. Bibliography. 1981 edition. ix + 283pp. 5⅜ x 8½. 0-486-67632-3

CALCULUS OF VARIATIONS, Robert Weinstock. Basic introduction covering isoperimetric problems, theory of elasticity, quantum mechanics, electrostatics, etc. Exercises throughout. 326pp. 5⅜ x 8½. 0-486-63069-2

THE CONTINUUM: A CRITICAL EXAMINATION OF THE FOUNDATION OF ANALYSIS, Hermann Weyl. Classic of 20th-century foundational research deals with the conceptual problem posed by the continuum. 156pp. 5⅜ x 8½. 0-486-67982-9

CHALLENGING MATHEMATICAL PROBLEMS WITH ELEMENTARY SOLUTIONS, A. M. Yaglom and I. M. Yaglom. Over 170 challenging problems on probability theory, combinatorial analysis, points and lines, topology, convex polygons, many other topics. Solutions. Total of 445pp. 5⅜ x 8½. Two-vol. set.
Vol. I: 0-486-65536-9 Vol. II: 0-486-65537-7

INTRODUCTION TO PARTIAL DIFFERENTIAL EQUATIONS WITH APPLICATIONS, E. C. Zachmanoglou and Dale W. Thoe. Essentials of partial differential equations applied to common problems in engineering and the physical sciences. Problems and answers. 416pp. 5⅜ x 8½. 0-486-65251-3

THE THEORY OF GROUPS, Hans J. Zassenhaus. Well-written graduate-level text acquaints reader with group-theoretic methods and demonstrates their usefulness in mathematics. Axioms, the calculus of complexes, homomorphic mapping, *p*-group theory, more. 276pp. 5⅜ x 8½. 0-486-40922-8

Math–Decision Theory, Statistics, Probability

ELEMENTARY DECISION THEORY, Herman Chernoff and Lincoln E. Moses. Clear introduction to statistics and statistical theory covers data processing, probability and random variables, testing hypotheses, much more. Exercises. 364pp. 5⅜ x 8½. 0-486-65218-1

STATISTICS MANUAL, Edwin L. Crow et al. Comprehensive, practical collection of classical and modern methods prepared by U.S. Naval Ordnance Test Station. Stress on use. Basics of statistics assumed. 288pp. 5⅜ x 8½. 0-486-60599-X

SOME THEORY OF SAMPLING, William Edwards Deming. Analysis of the problems, theory and design of sampling techniques for social scientists, industrial managers and others who find statistics important at work. 61 tables. 90 figures. xvii +602pp. 5⅜ x 8½. 0-486-64684-X

LINEAR PROGRAMMING AND ECONOMIC ANALYSIS, Robert Dorfman, Paul A. Samuelson and Robert M. Solow. First comprehensive treatment of linear programming in standard economic analysis. Game theory, modern welfare economics, Leontief input-output, more. 525pp. 5⅜ x 8½. 0-486-65491-5

PROBABILITY: AN INTRODUCTION, Samuel Goldberg. Excellent basic text covers set theory, probability theory for finite sample spaces, binomial theorem, much more. 360 problems. Bibliographies. 322pp. 5⅜ x 8½. 0-486-65252-1

GAMES AND DECISIONS: INTRODUCTION AND CRITICAL SURVEY, R. Duncan Luce and Howard Raiffa. Superb nontechnical introduction to game theory, primarily applied to social sciences. Utility theory, zero-sum games, n-person games, decision-making, much more. Bibliography. 509pp. 5⅜ x 8½. 0-486-65943-7

INTRODUCTION TO THE THEORY OF GAMES, J. C. C. McKinsey. This comprehensive overview of the mathematical theory of games illustrates applications to situations involving conflicts of interest, including economic, social, political, and military contexts. Appropriate for advanced undergraduate and graduate courses; advanced calculus a prerequisite. 1952 ed. x+372pp. 5⅜ x 8½. 0-486-42811-7

FIFTY CHALLENGING PROBLEMS IN PROBABILITY WITH SOLUTIONS, Frederick Mosteller. Remarkable puzzlers, graded in difficulty, illustrate elementary and advanced aspects of probability. Detailed solutions. 88pp. 5⅜ x 8½. 65355-2

PROBABILITY THEORY: A CONCISE COURSE, Y. A. Rozanov. Highly readable, self-contained introduction covers combination of events, dependent events, Bernoulli trials, etc. 148pp. 5⅜ x 8¼. 0-486-63544-9

STATISTICAL METHOD FROM THE VIEWPOINT OF QUALITY CONTROL, Walter A. Shewhart. Important text explains regulation of variables, uses of statistical control to achieve quality control in industry, agriculture, other areas. 192pp. 5⅜ x 8½. 0-486-65232-7

Math–Geometry and Topology

ELEMENTARY CONCEPTS OF TOPOLOGY, Paul Alexandroff. Elegant, intuitive approach to topology from set-theoretic topology to Betti groups; how concepts of topology are useful in math and physics. 25 figures. 57pp. 5⅜ x 8½. 0-486-60747-X

COMBINATORIAL TOPOLOGY, P. S. Alexandrov. Clearly written, well-organized, three-part text begins by dealing with certain classic problems without using the formal techniques of homology theory and advances to the central concept, the Betti groups. Numerous detailed examples. 654pp. 5⅜ x 8½. 0-486-40179-0

EXPERIMENTS IN TOPOLOGY, Stephen Barr. Classic, lively explanation of one of the byways of mathematics. Klein bottles, Moebius strips, projective planes, map coloring, problem of the Koenigsberg bridges, much more, described with clarity and wit. 43 figures. 210pp. 5⅜ x 8½. 0-486-25933-1

THE GEOMETRY OF RENÉ DESCARTES, René Descartes. The great work founded analytical geometry. Original French text, Descartes's own diagrams, together with definitive Smith-Latham translation. 244pp. 5⅜ x 8½. 0-486-60068-8

EUCLIDEAN GEOMETRY AND TRANSFORMATIONS, Clayton W. Dodge. This introduction to Euclidean geometry emphasizes transformations, particularly isometries and similarities. Suitable for undergraduate courses, it includes numerous examples, many with detailed answers. 1972 ed. viii+296pp. 6⅛ x 9¼. 0-486-43476-1

PRACTICAL CONIC SECTIONS: THE GEOMETRIC PROPERTIES OF ELLIPSES, PARABOLAS AND HYPERBOLAS, J. W. Downs. This text shows how to create ellipses, parabolas, and hyperbolas. It also presents historical background on their ancient origins and describes the reflective properties and roles of curves in design applications. 1993 ed. 98 figures. xii+100pp. 6½ x 9¼. 0-486-42876-1

THE THIRTEEN BOOKS OF EUCLID'S ELEMENTS, translated with introduction and commentary by Sir Thomas L. Heath. Definitive edition. Textual and linguistic notes, mathematical analysis. 2,500 years of critical commentary. Unabridged. 1,414pp. 5⅜ x 8½.. Three-vol. set.
Vol. I: 0-486-60088-2 Vol. II: 0-486-60089-0 Vol. III: 0-486-60090-4

SPACE AND GEOMETRY: IN THE LIGHT OF PHYSIOLOGICAL, PSYCHOLOGICAL AND PHYSICAL INQUIRY, Ernst Mach. Three essays by an eminent philosopher and scientist explore the nature, origin, and development of our concepts of space, with a distinctness and precision suitable for undergraduate students and other readers. 1906 ed. vi+148pp. 5⅜ x 8½. 0-486-43909-7

GEOMETRY OF COMPLEX NUMBERS, Hans Schwerdtfeger. Illuminating, widely praised book on analytic geometry of circles, the Moebius transformation, and two-dimensional non-Euclidean geometries. 200pp. 5⅝ x 8¼. 0-486-63830-8

DIFFERENTIAL GEOMETRY, Heinrich W. Guggenheimer. Local differential geometry as an application of advanced calculus and linear algebra. Curvature, transformation groups, surfaces, more. Exercises. 62 figures. 378pp. 5⅜ x 8½. 0-486-63433-7

History of Math

THE WORKS OF ARCHIMEDES, Archimedes (T. L. Heath, ed.). Topics include the famous problems of the ratio of the areas of a cylinder and an inscribed sphere; the measurement of a circle; the properties of conoids, spheroids, and spirals; and the quadrature of the parabola. Informative introduction. clxxxvi+326pp. 5⅜ x 8½. 0-486-42084-1

A SHORT ACCOUNT OF THE HISTORY OF MATHEMATICS, W. W. Rouse Ball. One of clearest, most authoritative surveys from the Egyptians and Phoenicians through 19th-century figures such as Grassman, Galois, Riemann. Fourth edition. 522pp. 5⅜ x 8½. 0-486-20630-0

THE HISTORY OF THE CALCULUS AND ITS CONCEPTUAL DEVELOPMENT, Carl B. Boyer. Origins in antiquity, medieval contributions, work of Newton, Leibniz, rigorous formulation. Treatment is verbal. 346pp. 5⅜ x 8½. 0-486-60509-4

THE HISTORICAL ROOTS OF ELEMENTARY MATHEMATICS, Lucas N. H. Bunt, Phillip S. Jones, and Jack D. Bedient. Fundamental underpinnings of modern arithmetic, algebra, geometry and number systems derived from ancient civilizations. 320pp. 5⅜ x 8½. 0-486-25563-8

A HISTORY OF MATHEMATICAL NOTATIONS, Florian Cajori. This classic study notes the first appearance of a mathematical symbol and its origin, the competition it encountered, its spread among writers in different countries, its rise to popularity, its eventual decline or ultimate survival. Original 1929 two-volume edition presented here in one volume. xxviii+820pp. 5⅜ x 8½. 0-486-67766-4

GAMES, GODS & GAMBLING: A HISTORY OF PROBABILITY AND STATISTICAL IDEAS, F. N. David. Episodes from the lives of Galileo, Fermat, Pascal, and others illustrate this fascinating account of the roots of mathematics. Features thought-provoking references to classics, archaeology, biography, poetry. 1962 edition. 304pp. 5⅜ x 8½. (Available in U.S. only.) 0-486-40023-9

OF MEN AND NUMBERS: THE STORY OF THE GREAT MATHEMATICIANS, Jane Muir. Fascinating accounts of the lives and accomplishments of history's greatest mathematical minds–Pythagoras, Descartes, Euler, Pascal, Cantor, many more. Anecdotal, illuminating. 30 diagrams. Bibliography. 256pp. 5⅜ x 8½. 0-486-28973-7

HISTORY OF MATHEMATICS, David E. Smith. Nontechnical survey from ancient Greece and Orient to late 19th century; evolution of arithmetic, geometry, trigonometry, calculating devices, algebra, the calculus. 362 illustrations. 1,355pp. 5⅜ x 8½. Two-vol. set. Vol. I: 0-486-20429-4 Vol. II: 0-486-20430-8

A CONCISE HISTORY OF MATHEMATICS, Dirk J. Struik. The best brief history of mathematics. Stresses origins and covers every major figure from ancient Near East to 19th century. 41 illustrations. 195pp. 5⅜ x 8½. 0-486-60255-9

Physics

OPTICAL RESONANCE AND TWO-LEVEL ATOMS, L. Allen and J. H. Eberly. Clear, comprehensive introduction to basic principles behind all quantum optical resonance phenomena. 53 illustrations. Preface. Index. 256pp. 5⅜ x 8½. 0-486-65533-4

QUANTUM THEORY, David Bohm. This advanced undergraduate-level text presents the quantum theory in terms of qualitative and imaginative concepts, followed by specific applications worked out in mathematical detail. Preface. Index. 655pp. 5⅜ x 8½. 0-486-65969-0

ATOMIC PHYSICS (8th EDITION), Max Born. Nobel laureate's lucid treatment of kinetic theory of gases, elementary particles, nuclear atom, wave-corpuscles, atomic structure and spectral lines, much more. Over 40 appendices, bibliography. 495pp. 5⅜ x 8½. 0-486-65984-4

A SOPHISTICATE'S PRIMER OF RELATIVITY, P. W. Bridgman. Geared toward readers already acquainted with special relativity, this book transcends the view of theory as a working tool to answer natural questions: What is a frame of reference? What is a "law of nature"? What is the role of the "observer"? Extensive treatment, written in terms accessible to those without a scientific background. 1983 ed. xlviii+172pp. 5⅜ x 8½. 0-486-42549-5

AN INTRODUCTION TO HAMILTONIAN OPTICS, H. A. Buchdahl. Detailed account of the Hamiltonian treatment of aberration theory in geometrical optics. Many classes of optical systems defined in terms of the symmetries they possess. Problems with detailed solutions. 1970 edition. xv + 360pp. 5⅜ x 8½. 0-486-67597-1

PRIMER OF QUANTUM MECHANICS, Marvin Chester. Introductory text examines the classical quantum bead on a track: its state and representations; operator eigenvalues; harmonic oscillator and bound bead in a symmetric force field; and bead in a spherical shell. Other topics include spin, matrices, and the structure of quantum mechanics; the simplest atom; indistinguishable particles; and stationary-state perturbation theory. 1992 ed. xiv+314pp. 6⅛ x 9¼. 0-486-42878-8

LECTURES ON QUANTUM MECHANICS, Paul A. M. Dirac. Four concise, brilliant lectures on mathematical methods in quantum mechanics from Nobel Prize-winning quantum pioneer build on idea of visualizing quantum theory through the use of classical mechanics. 96pp. 5⅜ x 8½. 0-486-41713-1

THIRTY YEARS THAT SHOOK PHYSICS: THE STORY OF QUANTUM THEORY, George Gamow. Lucid, accessible introduction to influential theory of energy and matter. Careful explanations of Dirac's anti-particles, Bohr's model of the atom, much more. 12 plates. Numerous drawings. 240pp. 5⅜ x 8½. 0-486-24895-X

ELECTRONIC STRUCTURE AND THE PROPERTIES OF SOLIDS: THE PHYSICS OF THE CHEMICAL BOND, Walter A. Harrison. Innovative text offers basic understanding of the electronic structure of covalent and ionic solids, simple metals, transition metals and their compounds. Problems. 1980 edition. 582pp. 6⅛ x 9¼. 0-486-66021-4